National Maritime Museum of China

国家海洋博物馆

黄克力　主编

天津大学出版社
TIANJIN UNIVERSITY PRESS

图书在版编目（CIP）数据

国家海洋博物馆 / 黄克力主编. -- 天津 : 天津大学出版社, 2021.9
ISBN 978-7-5618-6648-1

Ⅰ. ①国… Ⅱ. ①黄… Ⅲ. ①海洋－博物馆－建筑设计－天津②海洋－博物馆－藏品－天津－图录 Ⅳ. ① TU242.5 ② P72-28

中国版本图书馆 CIP 数据核字 (2020) 第 043485 号

GUOJIA HAIYANG BOWUGUAN

策划编辑　韩振平
责任编辑　郭　颖
装帧设计　谷英卉

出版发行　天津大学出版社
地　　址　天津市卫津路 92 号天津大学内（邮编：300072）
电　　话　022-27403647
网　　址　www.tjupress.com
印　　刷　北京雅昌艺术印刷有限公司
经　　销　全国各地新华书店
开　　本　260mm×260mm
印　　张　18 2/3
字　　数　70 千
版　　次　2021 年 9 月第 1 版
印　　次　2021 年 9 月第 1 次
定　　价　188.00 元

《国家海洋博物馆》
编委会

序

蔚蓝的海洋，是我们这颗星球上生命的起源之处。她孕育了自然万物的壮丽奇伟，滋养了人类文明的鲜活色彩，也在长久的历史记忆里，勾连起地域、民族与邦国的跌宕往事。

对于拥有五千年文明的中华民族来说，海洋的意义又多了一层厚重与复杂。早在新石器时期，河姆渡的初民就向大海投去了探索的目光。唐宋时期，海上丝绸之路商贾往来，中国人对海洋的了解和利用已经领先于世界。有明一代，郑和下西洋的伟业不仅成就了世界现存最早的航海图集，其规模、航程与采用的航海技术也开创了人类航海史的先河，成为中华民族走向海洋的一次盛大和平之旅。

从刳木为舟的先民时代到云帆济海的盛世王朝，中国人走过了与海相知的数千年时光。直至近代，历史的脉络陡然改变，海权沉沦的晦暗岁月成为横亘在每个中国人心中的难言之痛，也伏隐和生发着关于家国社稷、重构海图的渴望。

所幸，即使在至黯的时光里，依然有无数有识之士用无畏的付出为我们守住了这辽阔的海疆。今天的中国，不仅是拥有960万平方千米土地的陆地大国，更是拥有1.8万千米大陆海岸线、11 000多个岛屿的海洋大国。今天的中国人，也比历史上任何时候都更加理解这蔚蓝国土的神圣意义。

众所周知，21世纪是海洋的世纪。今日之海洋，早已成为陆地之外人类生存和发展的第二空间。关心海洋、认识海洋、经略海洋，不仅是当下之世的因势利导，更是关乎未来的谋长虑远。正因如此，从“提高海洋资源开发能力，发展海洋经济，保护海洋生态环境，坚决维护国家海洋权益”的蔚蓝之梦到“坚持陆海统筹，发展海洋经济，建设海洋强国”的远景目标，中华民族向海而行、向海而兴的号角不断吹响。一个个举世瞩目的成就，生动诠释了“海洋强国”之路上的坚守与辉煌，也在时代的意义之外，又多了一份在更加广阔的视野中为全球海洋建设和海洋合作谋篇布局的胸怀与作为。

往事不可遗忘，成就亦应铭记。从海洋大国走向海洋强国，是中华民族伟大复兴之路上的关键一步，国家海洋博物馆的筹划与建设，则成为新时代蓝色传奇中的一个重要里程碑。

2007年9月，30位两院院士联名上书国务院，建议兴建国家海洋博物馆；

2010年4月，经国务院同意、国家发展和改革委员会批复，国家海洋博物馆项目正式落位天津市滨海新区；

2014年10月，国家海洋博物馆建设工作启动；

2018年12月，国家海洋博物馆项目基本建成；

2019年5月1日，国家海洋博物馆正式向公众开放试运营。

跨越十余年的筹划与建设过程，从一个侧面表明了国家海洋博物馆的重要价值。作为我国第一座国家级综合性海洋博物馆，国家海洋博物馆所传递的内涵，不仅关乎泱泱中华与海相伴的悠远历史和向海图强的不懈追索，更关乎宇宙、地球与生命的来龙去脉，关乎海洋与人类的宏大命题，关乎构筑人类海洋命运共同体的精彩未来。

因此，在国家海洋博物馆2.3万平方米的展陈面积中，我们力求通过多个维度的展示，勾勒出关于海洋的不同侧面——她是生命源起的蔚蓝摇篮，也是仍待探索的未竟之地；她是不可撼动的蓝色国土，也是沟通全球的天然通道；她是自然馈赠的丰饶宝库，也是亟须保护的生态系统……海洋的身份如此多变，需要我们用寰球的眼光看待，用谦虚的态度求索，用现代的视角审悉，用博大的胸襟筹划，这是中国作为负责任大国的使命担当，也是让海洋连通起全人类福祉的必由之路。

新的时代正乘风破浪而来，国家海洋博物馆的建设与开放适逢其时。何其有幸，我们是这伟大事业的亲历者与参与者，也是向世界、向公众展示和讲述海洋故事的追梦人。2021年，是国家海洋博物馆试运营圆满成功后的正式开放之年。值此契机，我们奉献出这本图册，希望能够成为国家海洋博物馆的一份献礼，在中华民族伟大复兴的今天，传递出属于中国与世界悠远动人而又充满希望的蓝色回响。

是为序。

国家海洋博物馆馆长 黄克力

Preface

The blue ocean is where life on our planet originated. She has nurtured magnificent nature, nourished diversified human civilization, and recorded the ups and downs of regions, nations and states throughout the long history.

For the Chinese nation with 5,000 years of civilization, the significance of the ocean is more profound and complicated. As early as the Neolithic period, the primitive people of Hemudu shifted their eyes towards the ocean for exploration. During the Tang and Song Dynasties, the merchants came and went along the Maritime Silk Road, and Chinese people were already ahead of the world in understanding and utilizing the ocean. In the Ming Dynasty, Zheng He's voyages to the west not only left the earliest nautical atlas in the world, but also pioneered the history of human navigation in terms of scale, voyage and navigation technology, making it a grand and peaceful journey of the Chinese nation to the ocean.

From the days when our ancestors used hollow wood as boat to the prosperous ages when our people sailed for the ocean, Chinese people have spent thousands of years exploring the ocean. When it came to modern times, the trends of history changed abruptly and Chinese people suffered from the pains that maritime rights were seized by foreign countries, but at the same time they also had the desire to reconstruct the marine territory for the country.

Fortunately, even in these dark days, there were still countless far-sighted people who guarded our vast territorial ocean with fearless dedication. Now China is not only a large land country with 9.6 million square kilometers, but also a large maritime country with 18,000 kilometers of continental coastline and more than 11,000 islands. Today, Chinese people understand the sacred significance of this blue territory better than any time in history.

As we all know, the 21st century is the century of ocean. Today's ocean has long become another space for human survival and development beyond the land. Caring for the ocean, understanding the ocean and managing the ocean can not only lead the present world, but also have far-reaching significance. As a result, from the blue dream of "enhancing our capacity for exploiting marine resources, developing the marine economy, protecting the marine ecological environment and resolutely safeguarding China's maritime rights and interests" to the long-term goal of "pursuing coordinated land and marine development, developing the marine economy and building China into a strong maritime power", the Chinese nation is moving towards the ocean, and the horn of prosperity is blowing continuously. A number of achievements that attract worldwide attention vividly explain the Chinese nation's persistence and glory on the road to a "maritime power". Besides boasting the significance of the times, we have also shown the mind and action of Chinese people to make planning for global marine construction and marine cooperation in a broader vision.

The past should not be forgotten, and the achievements should be remembered. Moving from a big maritime country to a strong maritime country is a key step on the road to the great rejuvenation of the Chinese nation, and the planning and construction of the National Maritime Museum of China has become an important milestone in the blue legend of the new era.

In September 2007, 30 academicians of the Chinese Academy of Sciences and the Chinese Academy of Engineering wrote a joint letter to the State Council, proposing to establish the National Maritime Museum of China.

In April 2010, with the consent of the State Council and the approval of the National Development and Reform Commission, the site of the National Maritime Museum of China was officially selected in Tianjin Binhai New Area.

In October 2014, the construction of the National Maritime Museum of China started.

In December 2018, the construction of the National Maritime Museum of China was basically completed.

On May 1st, 2019, the National Maritime Museum of China was officially opened to the public for trial operation.

The planning and construction process, which spanned more than 10 years, reflects the important value of the National Maritime Museum of China. As the first national comprehensive maritime museum in China, the connotation conveyed by the National Maritime Museum of China is not only about China's long-history friendship with the ocean and the unremitting pursuit for maritime power, but also about the history of the universe, the earth and the life, the grand proposition of the ocean and the mankind, and the wonderful future of building a maritime community with mankind.

Therefore, in the National Maritime Museum of China with an exhibition area of 23,000 square meters, we strive to outline different aspects of the ocean through multi-dimensional display—she is the blue cradle of life and the unfinished land to be explored; she is an unshakable blue territory and a natural channel to communicate with the whole world; she is a bountiful treasure of nature and an ecosystem that needs urgent protection... The ocean's identity is so varied that we need to look at it from a global perspective, explore it with humility, view it from a modern perspective and plan it with a broad mind. This is China's mission as a responsible power and also the way to connect the ocean with the well-being of all mankind.

A new era is coming through the wind and waves, and the construction and opening of the National Maritime Museum of China has come at a better time. How fortunate we are to be the witnesses and participants of this great project, and also dream-catchers who present and tell stories about the ocean to the world and the public. The year 2021 witnessed the official opening of the National Maritime Museum of China after a successful trial operation. We dedicate this photo album on this occasion, hoping that it can be a gift from the National Maritime Museum of China that conveys the long-lasting and hopeful blue dream of China and the world under the context of the great rejuvenation of the Chinese nation.

This is the preface.

Curator of National Maritime Museum of China

目录

●远航时代

永无止境的探索之旅

●乐享海博

文旅融合的魅力图景

●后记

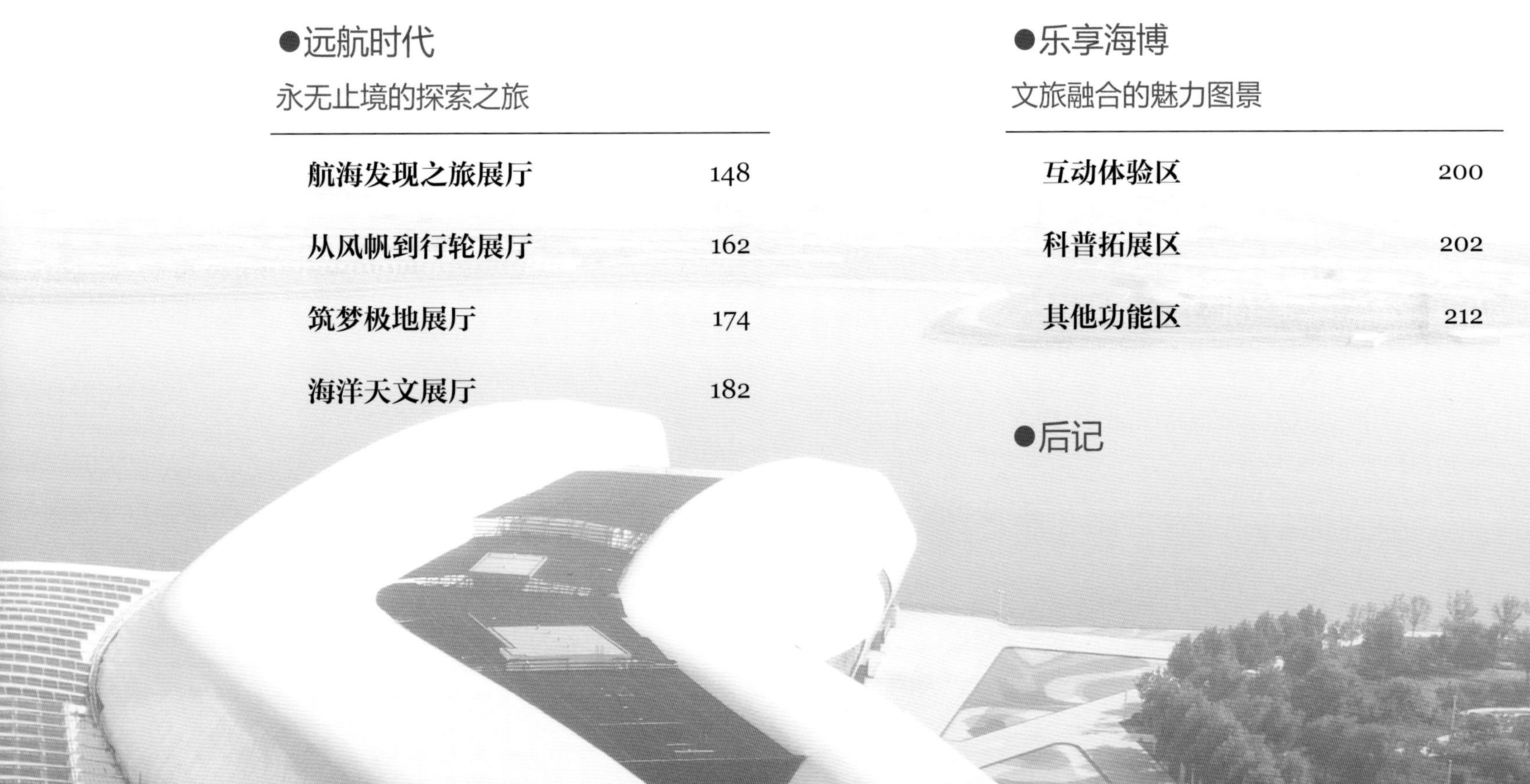

Contents

About the National Maritime Museum of China

The Only National Comprehensive Marine Museum in China

The Vicissitudes of the Ocean

The Ever-changing Marine Ecology

Sailing for the Ocean

Brilliant Marine Civilization of China

The Age of Oceangoing Voyage

A Never-ending Journey of Exploration

Enjoying Yourself in National Maritime Museum of China

Charming Pictures about Culture-tourism Integration

Postscript

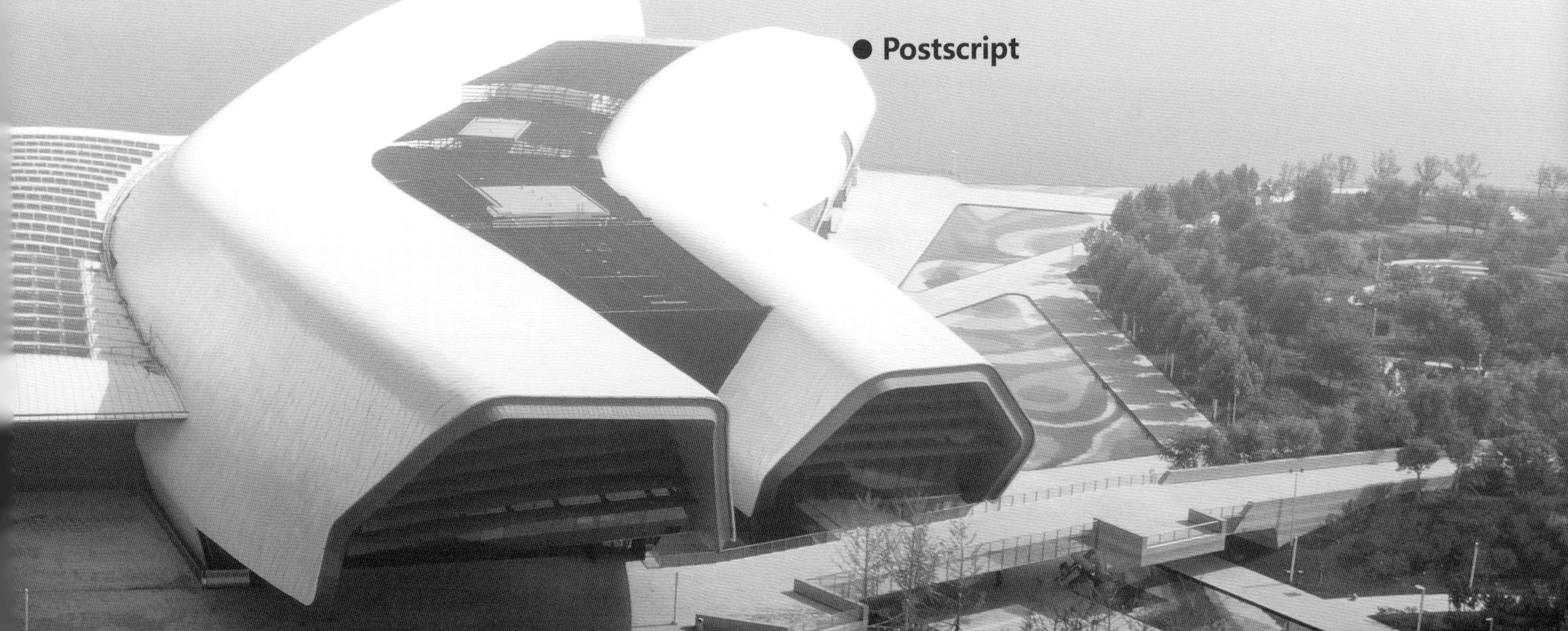

走进海博

我国唯一的国家级综合性海洋博物馆

About the National Maritime Museum of China

The Only National Comprehensive Marine Museum in China

国家海洋博物馆是经国务院同意、国家发展和改革委员会正式批复的国家重大项目,坐落于天津市滨海新区中新生态城滨海旅游区域，由国家海洋局与天津市人民政府共建。

2007年，30位两院院士联名上书时任国务院总理的温家宝同志，建议兴建国家海洋博物馆。2010年，经国务院同意、国家发展和改革委员会批复，项目正式落位滨海。2019年5月1日，国家海洋博物馆基本建设完成，启动试运营。

国家海洋博物馆展厅面积2.3万平方米，通过6大展区、16个展厅和3.6万余件珍贵藏品，全面展现了自然生态、历史人文、科学技术等多维视角下人类与海洋相互依存、和谐共生的密切关系。作为集收藏、展示、研究与教育于一体的我国唯一的国家级综合性海洋博物馆，国家海洋博物馆将服务我国海洋强国战略，为保护海洋生态文明，传承海洋历史文化，讲述当代海洋故事贡献力量。

The National Maritime Museum of China is a major national project consented by the State Council and officially approved by the National Development and Reform Commission. It is located in the Binhai tourism district of the Sino-Singapore Tianjin Eco-city in Tianjin Binhai New Area and is jointly built by the State Oceanic Administration and the Tianjin Municipal People's Government.

In 2007, 30 academicians of the Chinese Academy of Sciences and the Chinese Academy of Engineering wrote a joint letter to then Prime Minister Wen Jiabao, proposing to establish the National Maritime Museum of China. In 2010, with the consent of the State Council and the approval of the National Development and Reform Commission, the site of the project was officially selected in Tianjin Binhai New Area. On May 1st, 2019, the construction of the National Maritime Museum of China was basically completed, and started the trial operation.

The exhibition halls of the National Maritime Museum of China cover an area of 23,000 square meters, with six exhibition areas, 16 exhibition halls, and more than 36,000 precious collections. It comprehensively shows the close relationship of interdependence and harmonious coexistence between human beings and the ocean from a multi-dimensional perspective, such as natural ecology, history and culture, science and technology. As the only national comprehensive marine museum that integrates collection, exhibition, research and education, the National Maritime Museum of China will serve China's strategy to become a maritime power, contribute to the protection of marine ecological civilization, the inheritance of marine history and culture, and the narration of the contemporary ocean.

天津·国家海洋博物馆
NATIONAL MARITIME MUSEUM OF CHINA

沧海桑田

变化万千的海洋生态

The Vicissitudes of the Ocean

The Ever-changing Marine Ecology

当我们的目光遇见海洋，
就注定了这场关于沧海桑田的相遇。
从远古海洋到今日海洋，
从龙的时代到蓝色家园，
凝视海洋的过去与现在，
每件展品都在述说这颗蔚蓝星球的生命奇迹，
为我们的地球故事增加一个个瑰丽、神奇的注脚。

When we see the ocean,
we will also encounter the vicissitudes of the ocean.
From the ancient ocean to the today's ocean,
from the age of dragons to our blue homeland,
gazing at the past and present of the ocean,
each exhibit is telling about the miracle of life on this blue planet,
adding a magnificent and magical footnote to the story of our planet.

寒武纪
生命创新的时代
三叶虫
协同创新的
寒武纪海洋

远古海洋
展厅
Ancient Ocean Exhibition Hall

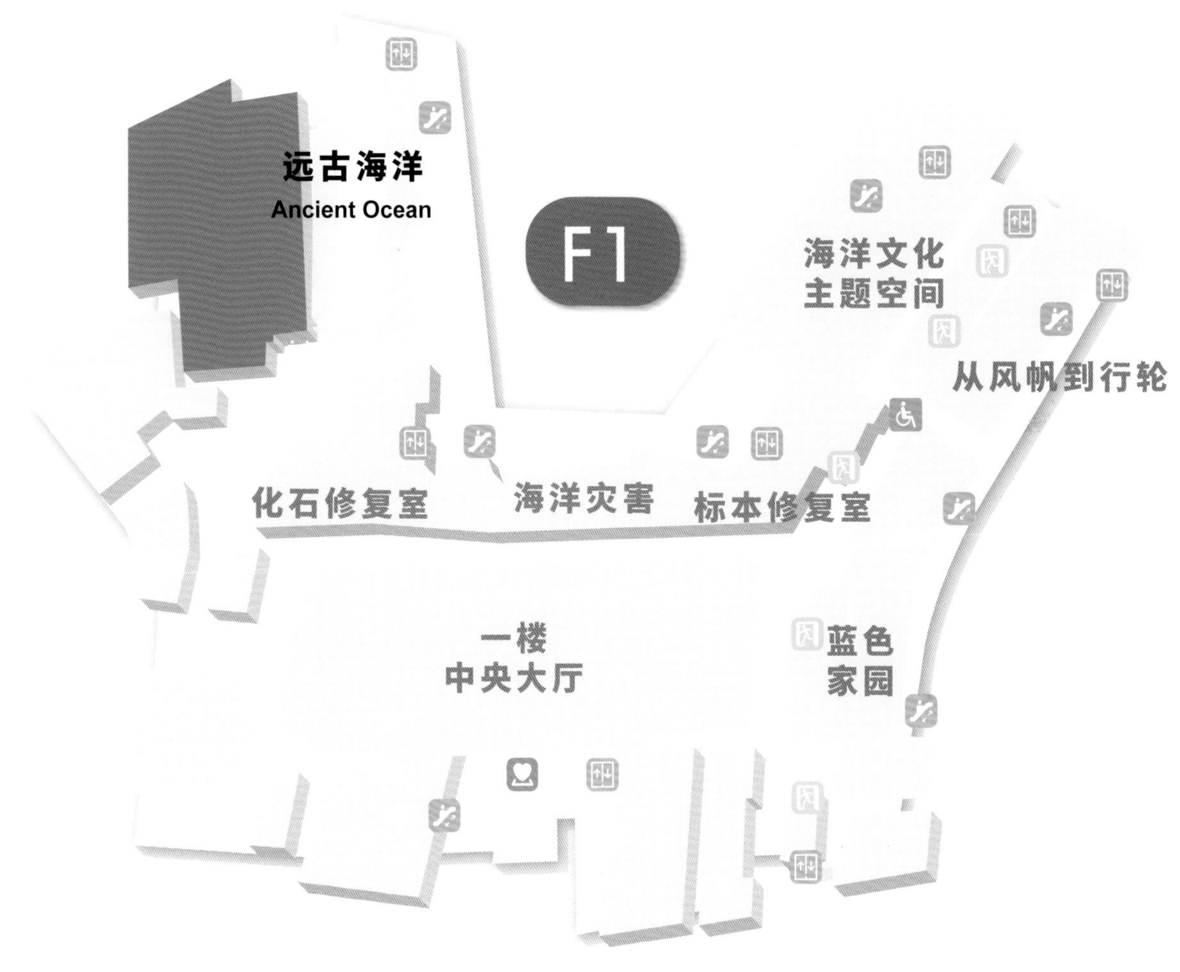

远古海洋展厅面积约为2 577平方米，主要展示了距今约40亿年以前，海洋形成之后，地球、海洋与生命波澜壮阔的演化过程，分为序厅、洪荒海洋、生命海洋、龙海沧桑、新生海洋五部分。

The Ancient Ocean Exhibition Hall covers an area of about 2,577 square meters, mainly displaying the magnificent evolution of the earth, ocean and life after the formation of the ocean about 4 billion years ago. It is divided into five parts: Preface Hall, Ancient Ocean, Ocean Life, Vicissitudes of Ocean, and New Look of Ocean.

序厅 / Preface Hall

序厅顶部以同心圆模拟宇宙星球运行的轨迹，当参观者仰望时，其倒影就成为浩渺宇宙中的一颗行星，映照出生命的奇迹与无数可能。

At the top of the Preface Hall, the concentric circles simulate the orbits of planets in the universe. When visitors look up, their reflection becomes a planet in the vast universe, reflecting the miracle of life and its countless possibilities.

几乎所有现生生命的雏形都
生命海洋
古生代海洋
LIFE OCEAN
PALEOZOIC OCEAN
ORIGIN
LIFE

寒武纪生命环形展示带

Cambrian Life Ring Display Zone

叠层石（中国天津） / Stromatolite (Tianjin, China)

菌藻类生物是地球孕育高级生命的基础，它们用数亿年的时光在天津市蓟州区累积了世界罕有的珍贵叠层石剖面。

Bacteria and algae, the basis for nurturing advanced life on the earth, have spent hundreds of millions of years accumulating rare and precious stromatolite profiles in the Jizhou District of Tianjin.

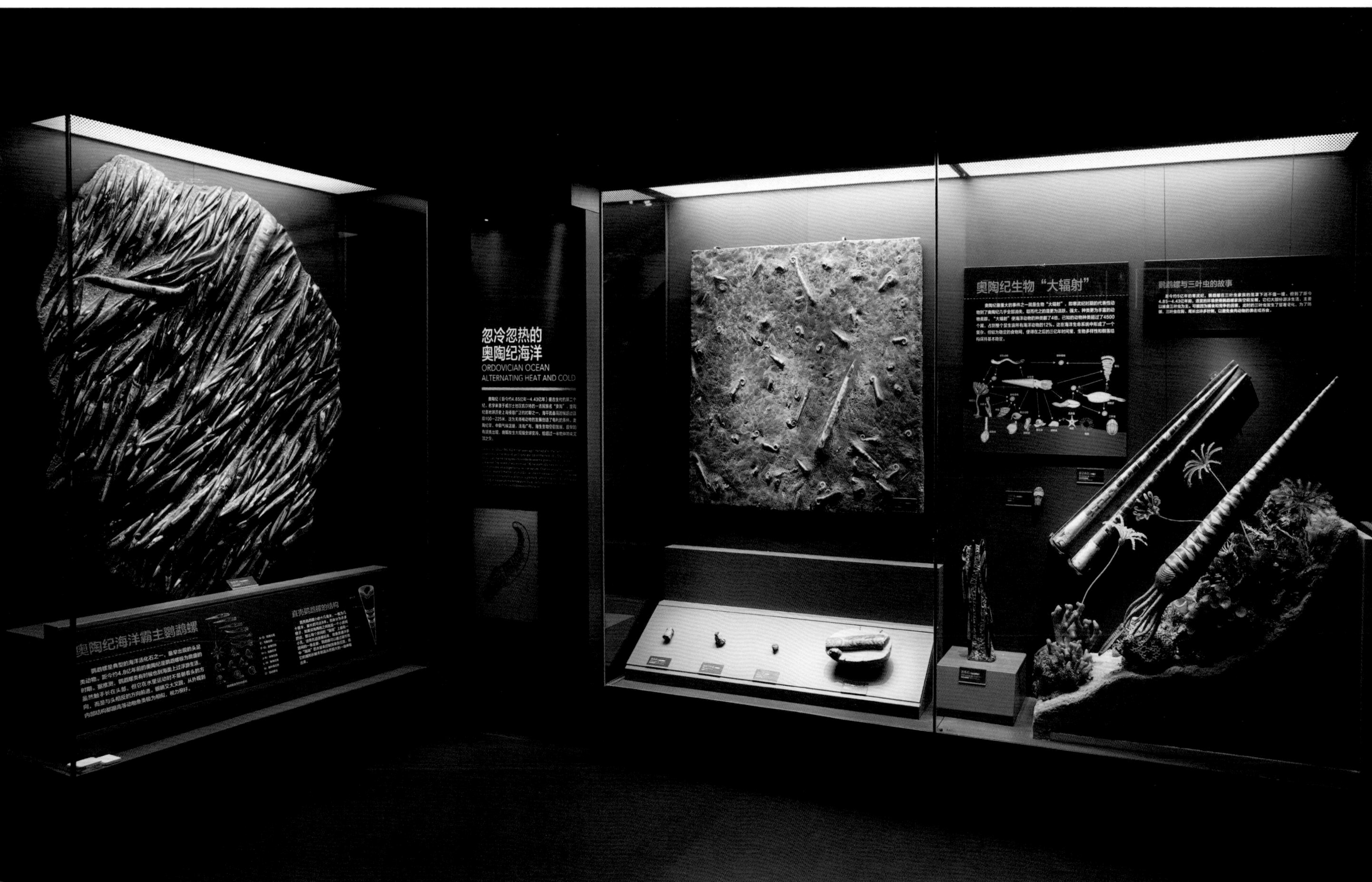
奥陶纪海洋霸主鹦鹉螺
忽冷忽热的
奥陶纪海洋
ORDOVICIAN OCEAN
ALTERNATING HEAT AND COLD
奥陶纪生物“大辐射”
鹦鹉螺与三叶虫的故事

菊石化石（摩洛哥）

Ammonite Fossil (Morocco)

沧海沧桑
中生代海洋
SAURUS OCEAN
MESOZOIC OCEAN
中生代
海洋生态
MESOZOIC
海洋生命繁盛
OCEAN LI

菊石
AMMONITES
与恐龙一起灭绝的无脊椎动物
跌宕起伏的
二叠纪海洋

来自我国多个省区的海生“龙”化石，以群像的方式再现了古生代大灭绝之后，生命从复苏到繁盛的灿烂历程。横向对比的展陈方式、复古厚重的展示环境，是向我国北方第一座自然博物馆——北疆博物院的致敬。跨越时光的追溯，不仅是形式的尊重，也是科学精神的延续。

Marine "dragon" fossils discovered in several provinces and regions in China reproduce the splendid course of life from recovery to prosperity after the Paleozoic extinction in the form of group images. The horizontally contrasting exhibition style and retro-heavy display environment are a tribute to the Musée Hoangho-Paiho, the first natural museum in northern China. Tracing back across time is not only the respect to the form, but also the continuation of the scientific spirit.

THE WITNESS
生命复苏的见证
中国三叠纪海洋爬行动物群
TRIASSIC MARINE REPTILES BIOTA IN

盘县-罗平
动物群 距今约2.44亿年
THE PANXIAN AND LUOPING FAUNAS

龟山巢湖鱼龙

CHAOHUSAURUS FAUNA
安徽
巢湖龙动物群

龙鱼

食物网

南漳 远安动物群
距今约2.47亿年
NANZHANG YUAN'AN FAUNA

独一无二的爬行类动物之一
南漳湖北鳄

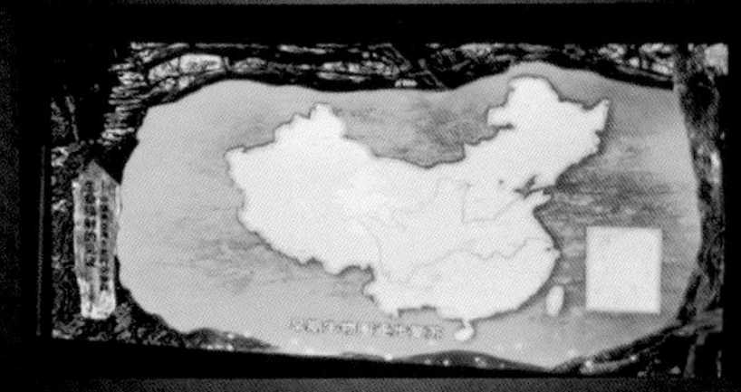

兴义
富源 动物群
幻龙
海龙类
楯齿龙

盘县-罗平
动物群 距今约2.44亿年
THE PANXIAN AND
LUOPING FAUNAS
龟山巢湖鱼龙
最原始的鱼龙之一
CHAOHUSAURUS
FAUNA
安徽
距今约2.48亿年
巢湖龙动物群
龙鱼
高度特化的辐鳍鱼类
罗平生物群 食物网
南漳 远安动物群
NANZHANG
YUAN'AN FAUNA
距今约2.47亿年
独一无二的爬行类动物之一
南漳湖北鳄

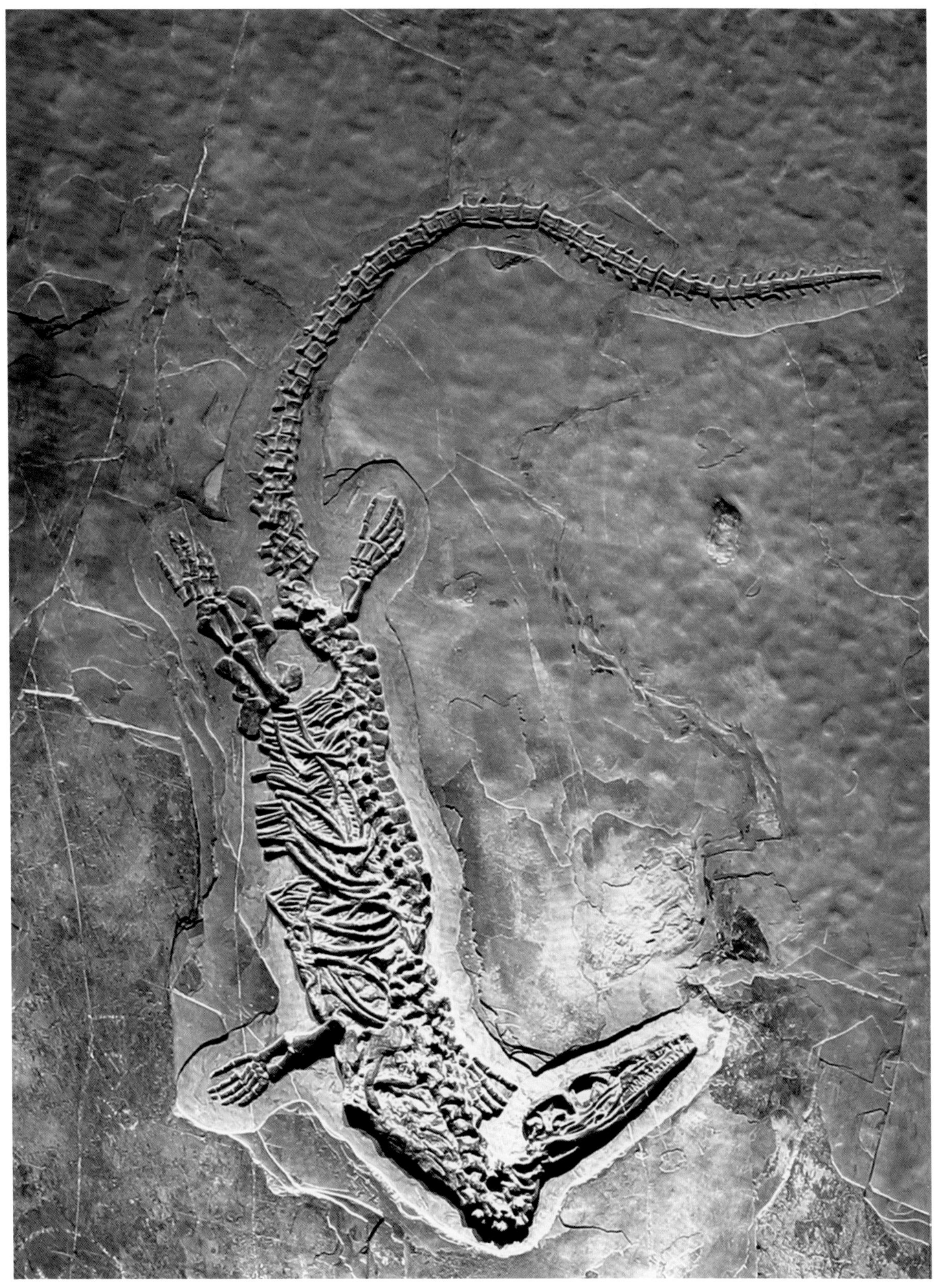

海龙（中国贵州兴义）

Syngnathus Schlegeli
(Xingyi, Guizhou Province, China)

杯椎鱼龙（中国贵州关岭）

Cymbospondylus(Guanling, Guizhou Province, China

三叠纪晚期海百合（中国贵州关岭）

Late Triassic Crinoids (Guanling, Guizhou Province, China)

贝壳堤（中国天津） / Shell Dike (Tianjin, China)

天津贝壳堤是世界著名的三大古贝壳堤之一，是古海岸变迁的珍贵海洋记忆。

Tianjin Shell Dike is one of the three world-famous ancient shell dikes, which is a precious marine memory of ancient coastal changes.

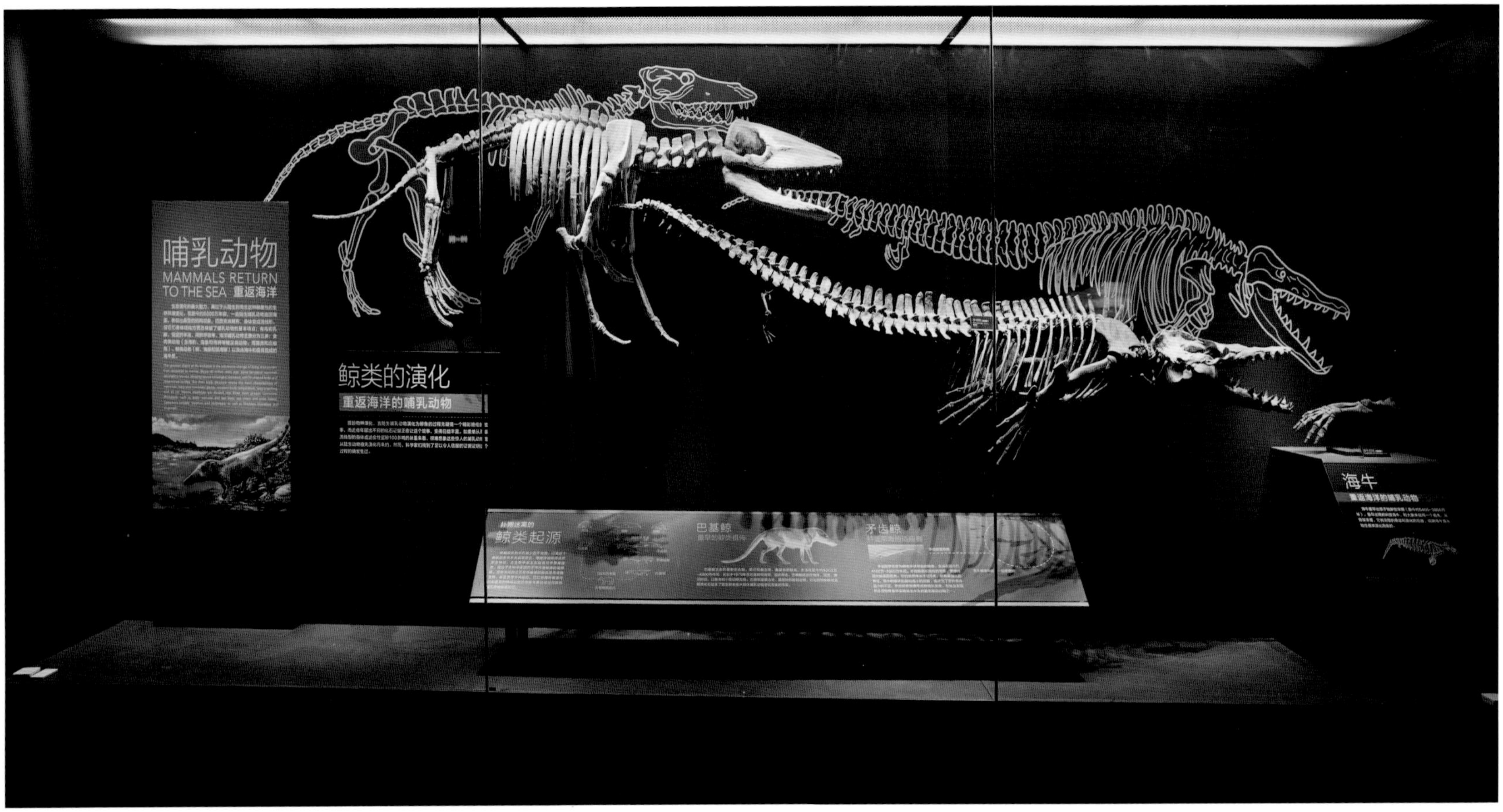
哺乳动物
MAMMALS RETURN
TO THE SEA 重返海洋
鲸类的演化
重返海洋的哺乳动物
鲸类起源
巴基鲸
海牛
重返海洋的哺乳动物

新生海洋
新生代海洋
NEWBORN OCEAN
CENOZOIC OCEAN
新生代
CENOZOIC MARINE ECOLOGY
海洋生态
新生代
海洋地理环境
CENOZOIC MARINE GEOGRAPHY
巨齿鲨
新生代典型地质事件

龙的时代
展厅
The Age of Dragons Exhibition Hall

龙的时代展厅面积约为1 000平方米，讲述了中生代生命从海洋诞生后出走陆地、占领天空，又回归海洋的故事。这场波澜壮阔的生命之旅，最终止步于白垩纪。但那个属于龙的时代，时至今日依然令人着迷。

The Age of Dragons Exhibition Hall covers an area of about 1,000 square meters and tells the story of Mesozoic life that was born from the ocean and then went out to the land, occupied the sky, and returned to the ocean at last. This magnificent journey of life ended in the Cretaceous. But that era, which belonged to dragons, is still fascinating today.

“龙”的时代
THE AGE OF DRAGONS
中生代的海洋、陆地和天空
OCEAN, LAND AND SKY OF THE MESOZOIC
关于“龙”和“海怪”
ABOUT DRAGON AND SEA MONSTER

在龙的时代，遍布海洋、陆地与天空的生命共同绘就了波澜壮阔的进化图景，也给今天的人们留下了无数神秘与未知。

In the age of dragons, life all over the sea, land and sky painted a magnificent picture of evolution together and left countless mysteries and unknowns to people today.

MOVEMENT
防御 DEFENSE

鱼龙化石（中国云贵地区）

Ichthyosaur Fossil (Yunnan and Guizhou Provinces, China)

国内最大的鱼龙化石，长达11米。

The largest ichthyosaur fossil in China is 11 meters long.

龙的时代终结于白垩纪，这些生机盎然的远古场景早已成为遥远的想象。但湮灭从不是生命的终点，而是更加宏大的开始。

The age of dragons ended in the Cretaceous, and these vibrant ancient scenes have long since become a distant imagination. But annihilation is never the end of life, but the beginning of something even grander.

今日海洋
展厅
Today's Ocean Exhibition Hall

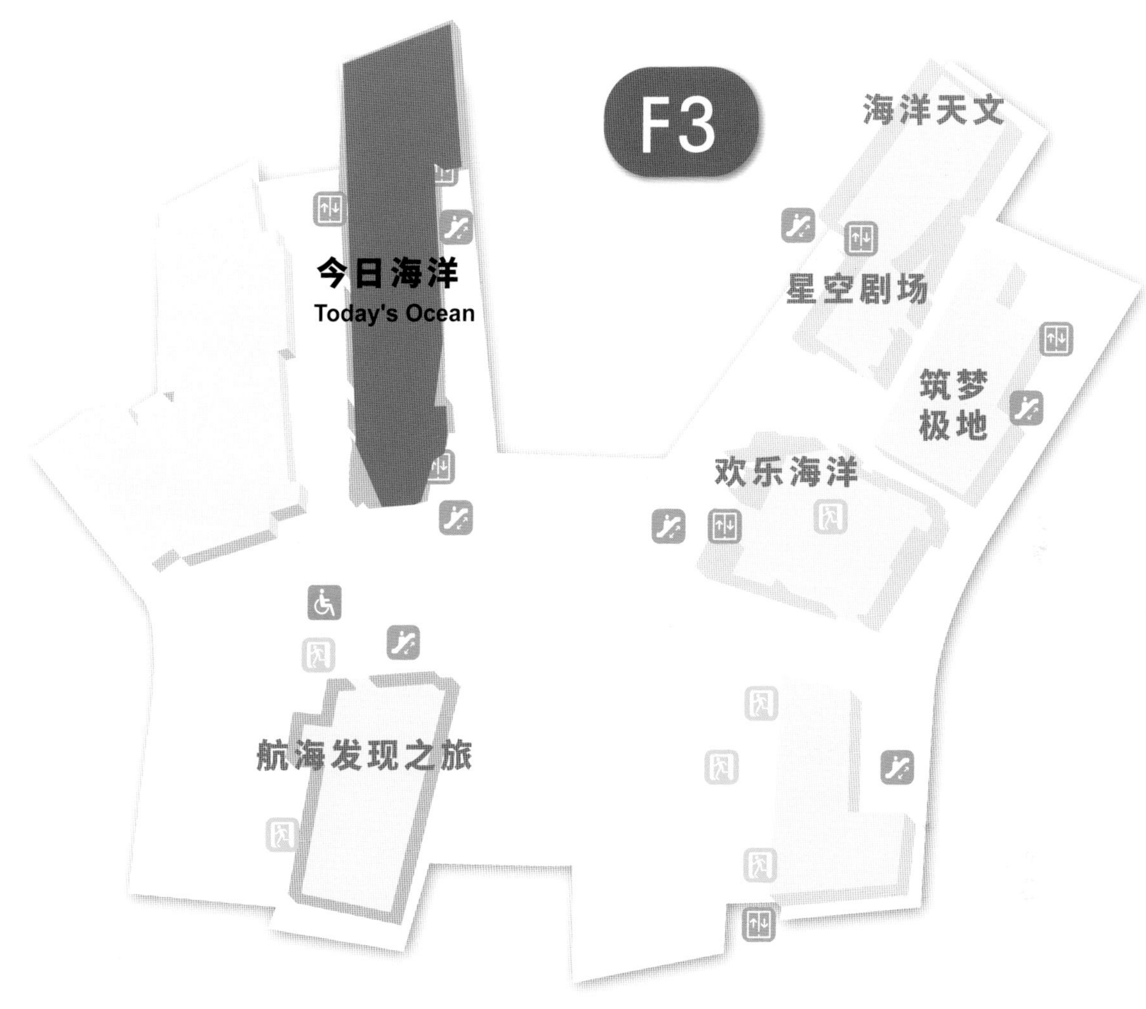

今日海洋展厅面积约为2 800平方米，通过5 000余件精美的标本，串联和讲述了以海洋为源点的现代自然生态，以“寰球同此凉热”的全球视野，呼唤人们为保护海洋、保护自然作出努力。

The Today's Ocean Exhibition Hall covers an area of about 2,800 square meters. Through more than 5,000 exquisite specimens, it connects and tells the modern natural ecology with the ocean as the source, and calls on people to make efforts to protect the ocean and nature with the global vision that "we share warm and cold in the world together".

解析海洋
海水
大洋中脊
大洋盆地
深入海洋
大陆边缘
海洋地理构成
海水中的有机物

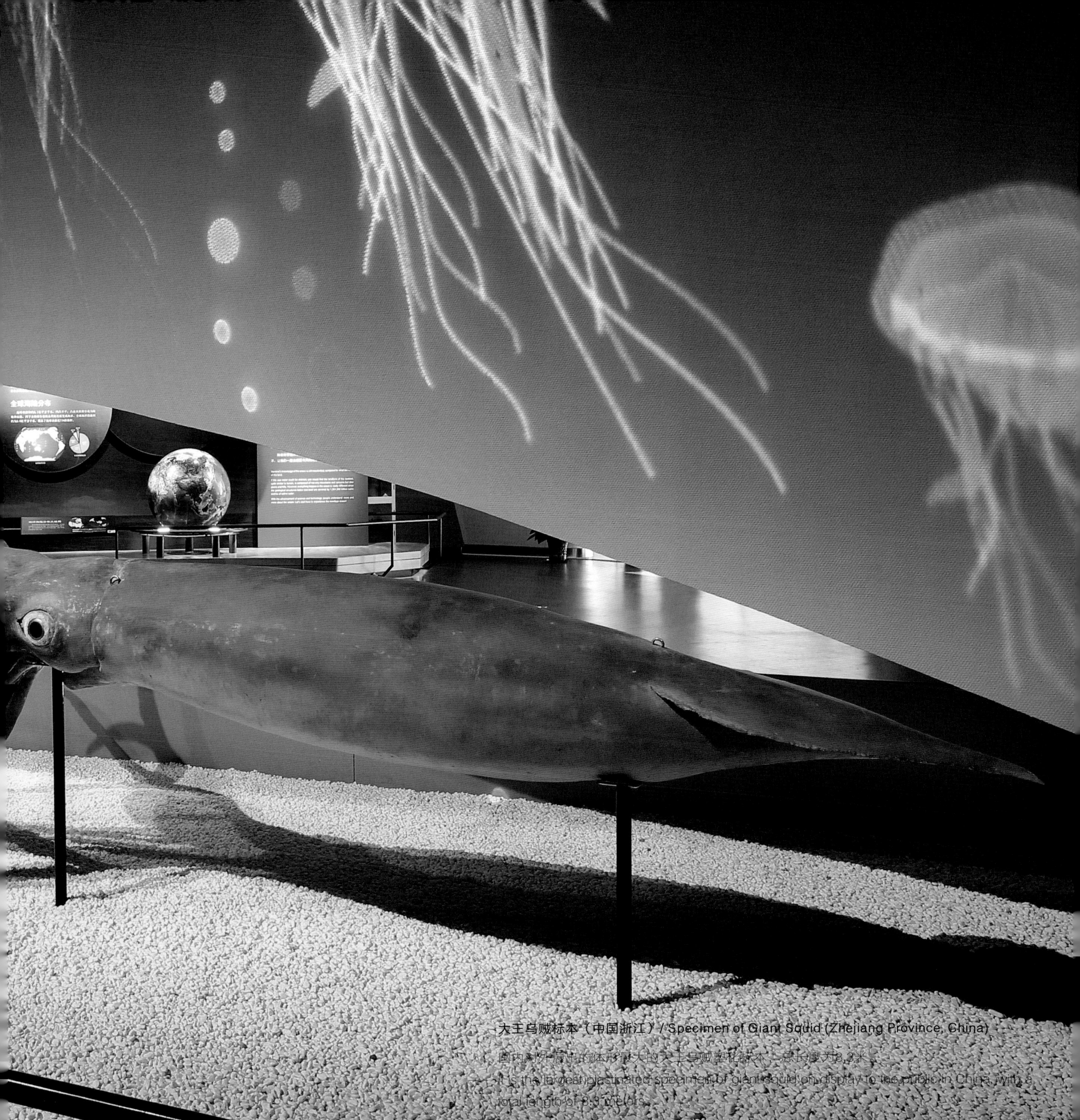

大王乌贼标本（中国浙江）/ Specimen of Giant Squid (Zhejiang Province, China)

国内对外展出的体形最大的大王乌贼塑化标本，总长度为8.3米。

It is the largest plastinated specimen of giant squid on display to the public in China, with a total length of 8.3 meters.

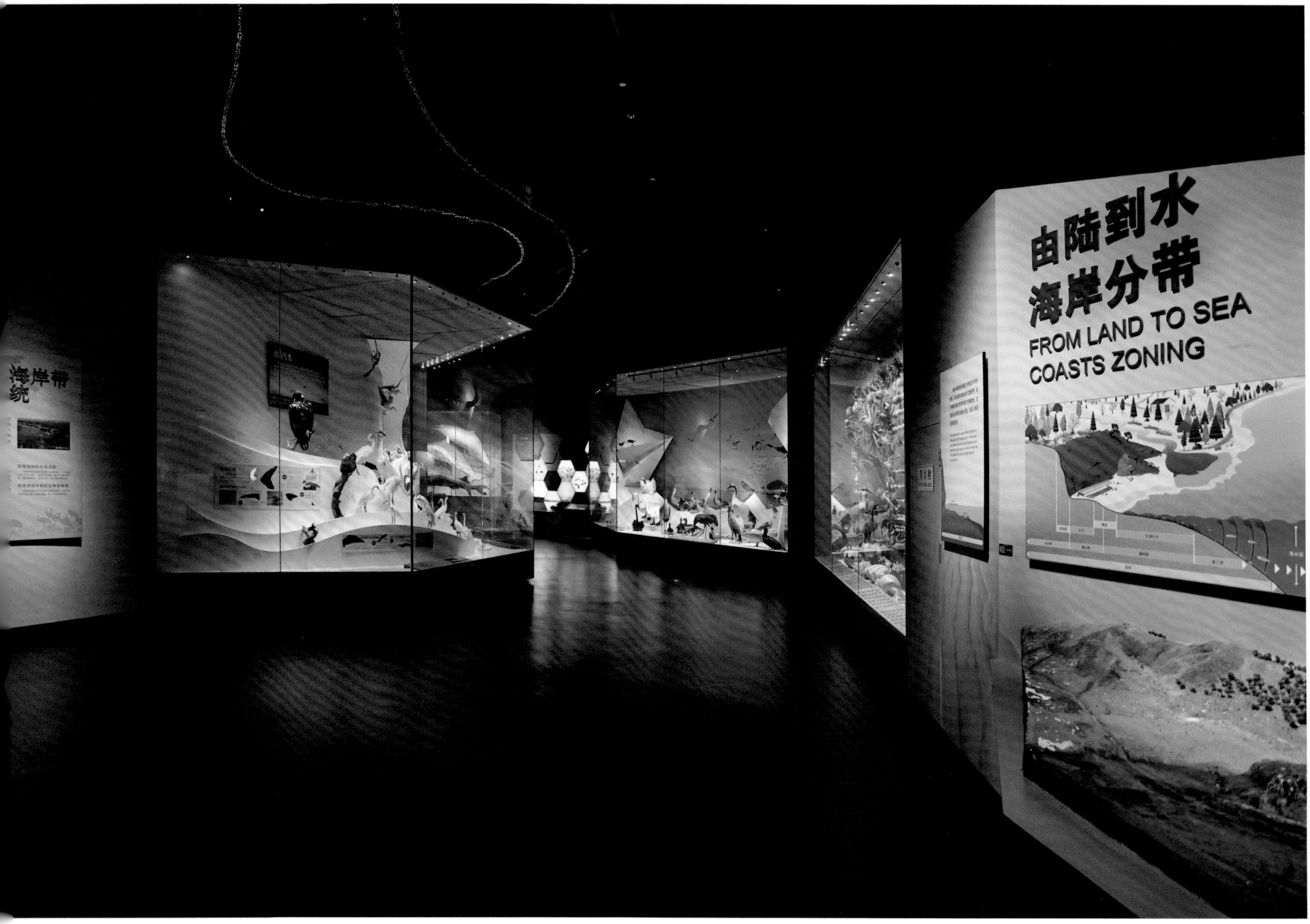
由陆到水
海岸分带
FROM LAND TO SEA
COASTS ZONING
海岸带
统

刺鲀
飞鱼

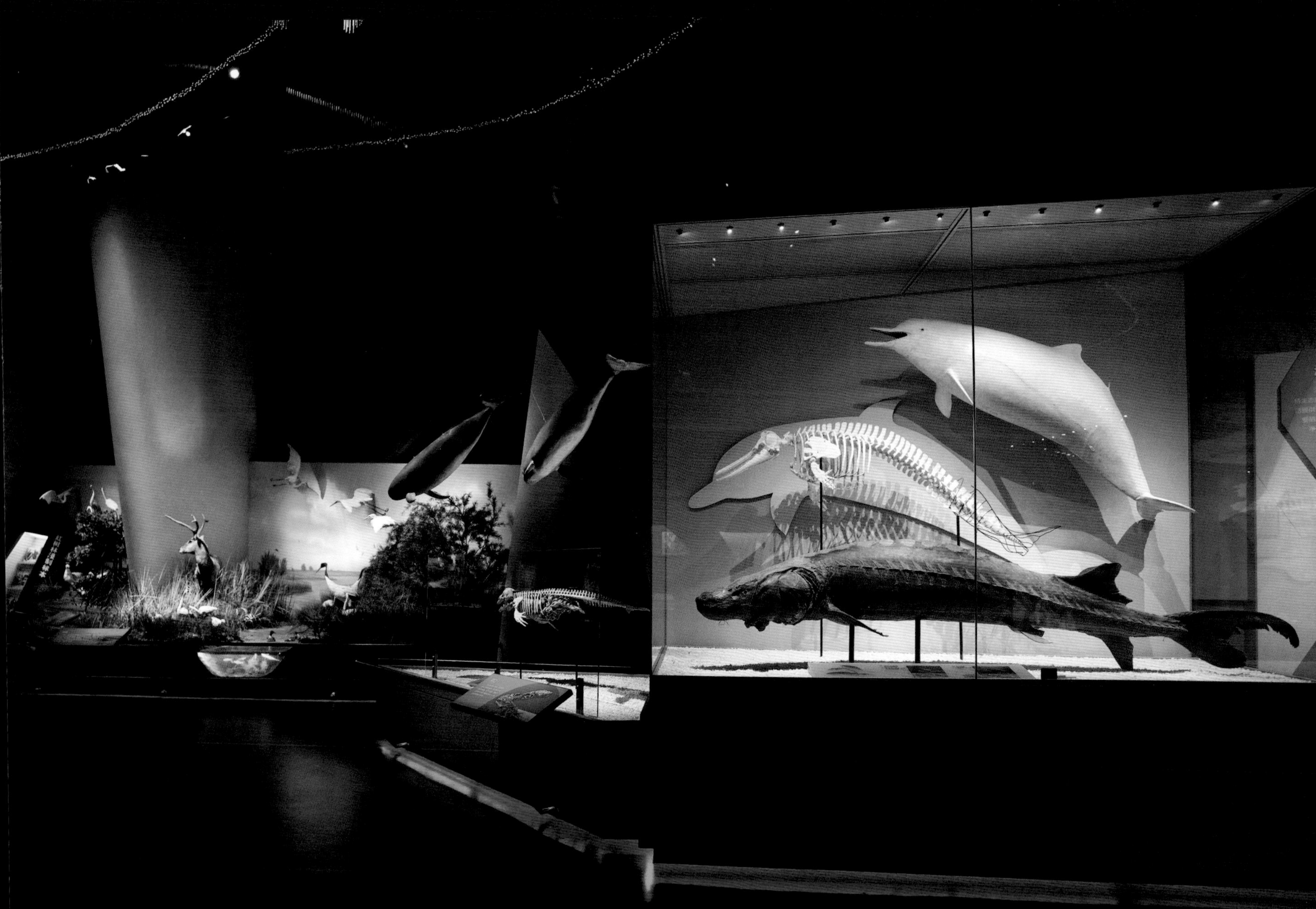

鲸鲨标本 / Whale Shark Specimen

标本长约9.4米，2016年发现于山东附近海域并由我馆实施保护性收藏，为近年来保存较完好、体形较大的鲸鲨馆藏。

The whale shark specimen of about 9.4 meters long was found in maritime area of Shandong Province in 2016 and collected by National Maritime Museum of China for protection, which is a well preserved museum collection with large body shape in recent years.

无光层

海洋生物家谱

珊瑚

小须鲸标本/ Specimen of Minke Whale

目前国内较大的海洋哺乳动物塑化标本，体长约6.2米。

It is a plastinated specimen of a large marine mammal in China, with a body length of about 6.2 meters.

单板纲
Monoplacophora
无板纲
Aplacophora
多板纲
Polyplacophora

贝类环形展示带 / Shellfish Ring Display Zone

藏有目前世界上较完整的一套中国海域贝类标本，共4 000余件。

The zone displays a relatively complete collection of shellfish specimens in Chinese waters in the world today, with more than 4,000 specimens.

保护海洋可持续发展
GUARD THE SUSTAINABLE DEVE OPMENT CHANGES OF OCEAN
生物栖息地保护
保护海洋
我们在行动
自然海洋环境变化
生物环境灾害
海底地质灾害
变化不仅仅是坏事！

蓝色家园
展厅

Blue Homeland Exhibition Hall

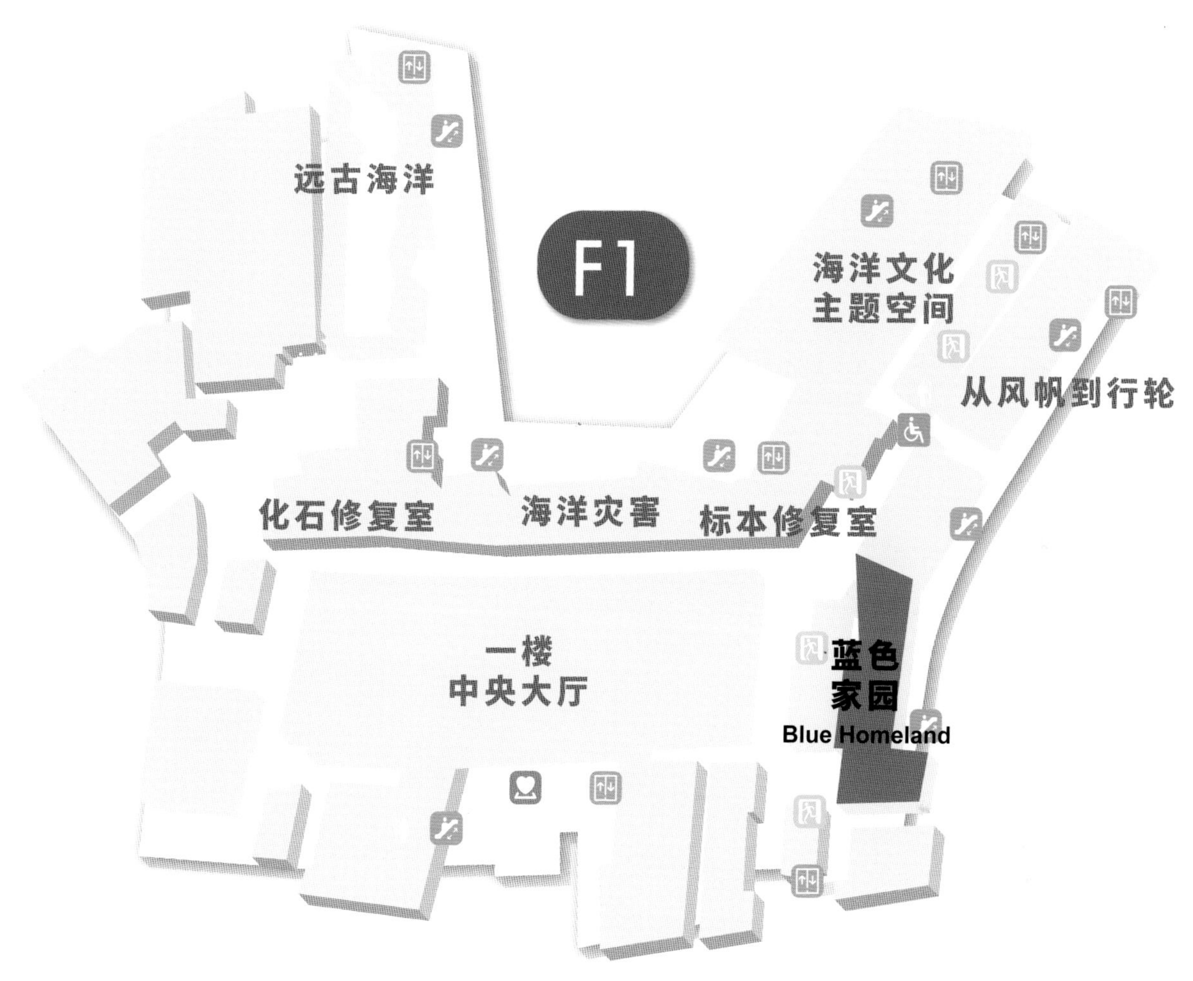

蓝色家园展厅为活体生物展厅。展览以走进红树林、探秘海藻场、遨游珊瑚礁、海洋猎手、幻彩水母、海底探奇为分区，以海洋生物的生存环境为线索，系统展示了广袤海洋空间的奇妙生态。

The Blue Homeland Exhibition Hall is a living organism exhibition hall. The exhibition is divided into mangroves, seaweed farms, coral reefs, marine hunters, colorful jellyfish, and undersea exploration, to systematically show the wonderful ecology of the vast ocean space by taking the living environment of marine organisms as a clue.

海洋灾害

体验厅

Marine Disaster Experience Hall

属于海洋的，不仅有绚烂奇美的自然生态，也有摧枯拉朽的骇人灾难。在1 850平方米的海洋灾害体验厅中，游客可以通过沉浸体验、复原场景、视频展示等多种形式了解海洋灾害和防灾减灾知识。

What belongs to the ocean includes not only gorgeous and exotic natural ecology, but also devastating and horrific disasters. In the 1,850-square-meter Marine Disaster Experience Hall, the visitors can learn about marine disasters and disaster prevention and mitigation through immersive experiences, restored scenes, video displays, etc.

生物
BIOSPHERE
岩石
水
冰冻
大气
ATMOSPHERE
地球系统概述
EARTH SYSTEM OVERVIEW

造浪池 / Wave-making Pool

造浪池长50米、宽3.5米、高2.3米，能够通过模拟多种海浪模式，直观展现巨浪的生成与破坏力。

The wave-making pool is 50 meters long, 3.5 meters wide and 2.3 meters high, which can visually show the generation and destructive power of huge waves by simulating a variety of wave patterns.

▽ +1.85m
胸墙
沉箱
▽ +1.50m
块石
块石
▽ +0.80m
块体
1:2
块体
▽ +0.41m
1:2
垫层
1:2
▽ +0.20m
护底
▽ 地平高程

海洋灾害剧场 / Marine Disaster Theater

以实景视频结合水雾喷淋、模拟强风等综合体感手段，实现台风呼啸的沉浸式体验。

The immersive experience of typhoon whistling is realized by means of the real-scene video combined with water spraying and strong wind simulation.

实验室

澳大利亚 Australia
墨西哥 Mexico
大西洋西岸 Western
生态灾害
ECOLOGICAL DISASTERS
几种常见生态灾害
SEVERAL COMMON ECOLOGICAL DISASTERS
生态灾害及其诱因
ECOLOGICAL DISASTERS AND THEIR CAUSES

云帆济海

中华海洋文明的华章

Sailing for the Ocean

Brilliant Marine Civilization of China

中华民族的悠远历史，从不缺少海洋的身影。
海洋收藏了浩瀚文明中文化往来的历史记忆，
承载了风雨晦暗里创巨痛深的民族之殇，
更见证了今日盛世繁华、大国崛起的不世伟业，
长风破浪会有时，直挂云帆济沧海。
我们对蔚蓝的追寻仍在继续，伟大的征程正逢其时。

We can find the presence of the ocean everywhere in the long history of the Chinese nation.
The ocean has collected the historical memories of cultural exchanges in the vast civilization,
carried the national grief and great pain in the darkest days,
and witnessed today's prosperity and the rise of the great nation.
A time will come to ride the wind and cleave the waves, and we will set our cloud-white sail and cross the sea which raves.
Our pursuit of a blue planet continues, and the great journey will come at a better time.

中华海洋文明
第一篇章
Marine Civilization of China 1

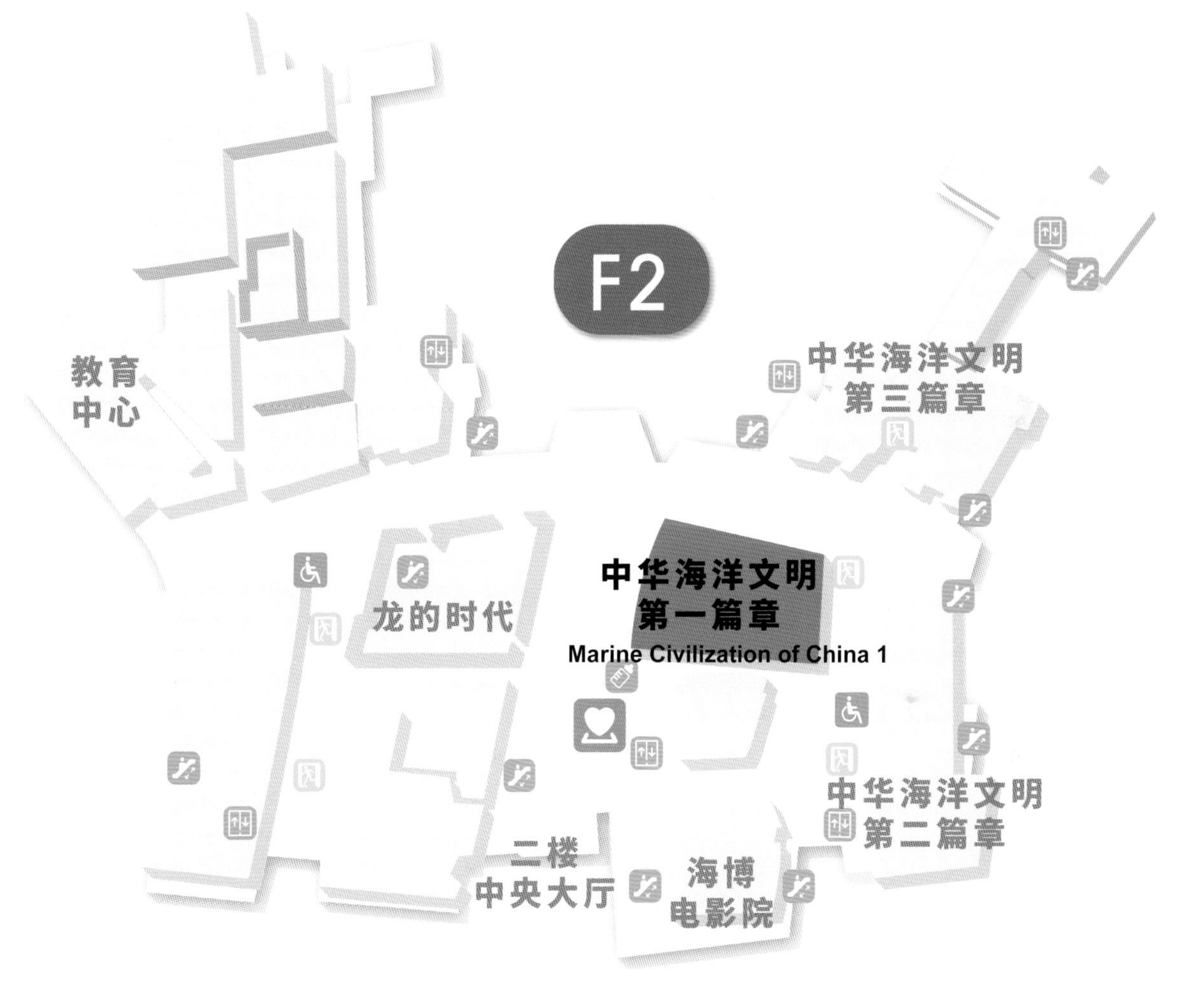

中华海洋文明第一篇章分为向海而生、陆海融汇、海疆经略、海上丝绸之路的繁荣、郑和下西洋五部分,全面展现了中华民族认识海洋、探索海洋、开发海洋的悠久历史。

Marine Civilization of China 1 is divided into five parts: "born to the sea", "land-sea integration", "strategies of coastal areas and territorial seas", "flourishing Maritime Silk Road" and "Zheng He's voyages to the west ", which comprehensively show the long history of the Chinese nation in understanding the sea, exploring the sea, and developing the sea.

独木舟（隋唐时期）

Dugout Canoe (Sui and Tang Dynasties)

该独木舟长13.8米，腹径约0.95米，发现于广东省西江流域。它是目前国内有记载的保存状况较好、长度较长的独木舟，对研究中国古代舟船发展史、航海史和造船史具有重要意义。

The dugout canoe, which is 13.8 meters long with a ventral diameter of about 0.95 meters, was discovered along Xijiang River basin, Guangdong Province. It is a well-preserved dugout canoe with a long length recorded in China at present, which is of great significance to the study of the ship development history, navigation history and shipbuilding history of ancient China.

唐宋时期海上丝绸之路文物

The cultural relics of Maritime Silk Road in Tang and Song Dynasties

海上丝绸之路是中国古代对外贸易和文化交往的海上通道，其萌芽于商周，发展于春秋战国，形成于秦汉，兴盛于唐宋，转变于明清，是已知最古老的海上航线之一，承载着中国人与海洋的不解之缘。

The Maritime Silk Road is a sea route for foreign trade and cultural exchanges in ancient China, which sprouted in Shang and Zhou dynasties, developed in the Spring and Autumn Period and the Warring States Period, formed in the Qin and Han Dynasties, flourished in the Tang and Song Dynasties, and transformed in the Ming and Qing Dynasties. It is one of the oldest known sea routes, carrying the inseparable relationship between the Chinese people and the ocean.

福船（现代复制）

Fu Sailing Ship (modern replicated model)

福船位于海洋文化主题空间，原型为宋元时期的远洋货运船，长27.8米、宽8.6米，再现了水密隔舱和鱼鳞式搭接两项中国古代造船重大发明，也从一个侧面展示了当时中国海上贸易的庞大规模与繁盛场面。

Located in the Marine Culture Theme Space, the prototype of the Fu Sailing Ship model is an oceangoing cargo ship in Song and Yuan Dynasties, with a length of 27.8 meters and a width of 8.6 meters. It reproduces two major inventions of ancient Chinese shipbuilding, namely watertight compartment and fish scale lap, and also reflects the huge scale and prosperous scene of China's maritime trade at that time.

中华海洋文明
第二篇章

Marine Civilization of China 2

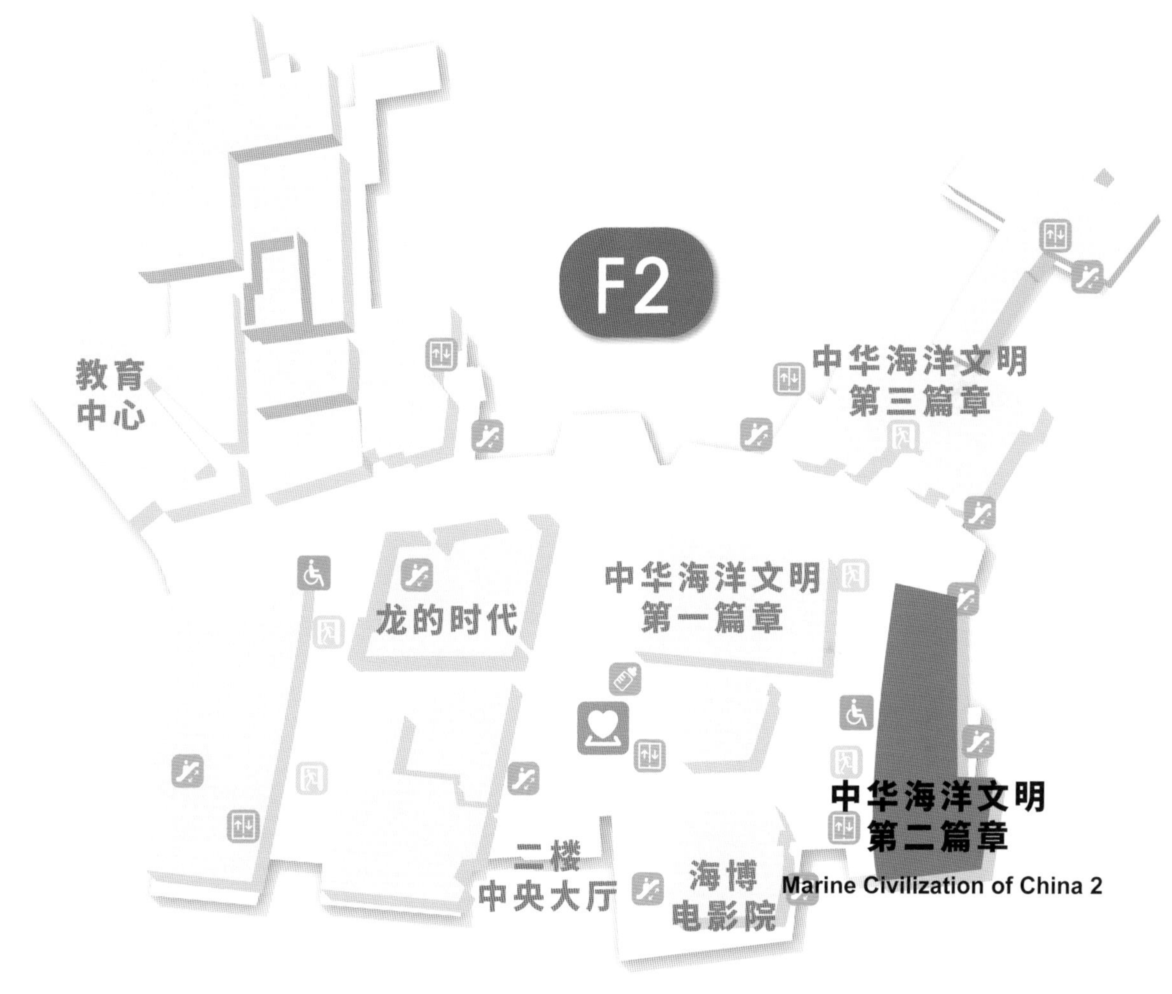

中华海洋文明第二篇章分为禁海与开海、郑成功与台湾、保守与被动开放、觉醒与探索四部分，揭示了明清以来中华民族探索海洋发展的近代化之路。

Marine Civilization of China 2 is divided into four parts: "maritime embargo and opening-up", "Zheng Chenggong and Taiwan", "isolation and passive opening-up" and "awakening and exploration", which reveal the Chinese nation's exploring the way of modernization of marine development since the Ming and Qing Dynasties.

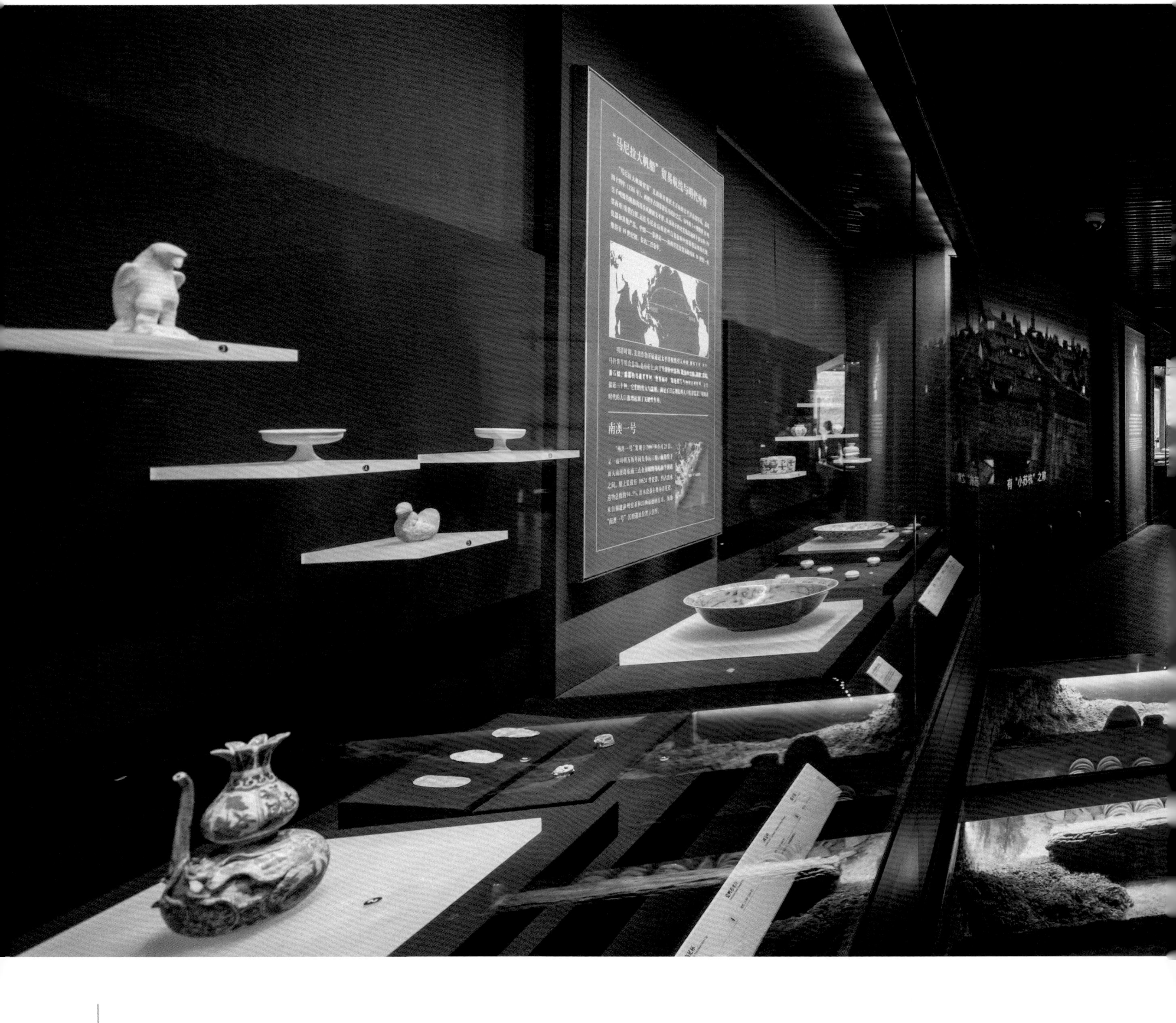
“马尼拉大帆船”贸易航线与明代外贸
南澳一号
有“小苏杭”之称

重出外洋
REOPENING-UP TO FOREIGN COUNTRIES
在禁海与开海的争论中，明代对外贸易与交往的步伐从未停滞。明代中期，民间海洋力量崛起，冲破海禁重出外洋。隆庆年间，月港（今漳州）开放，东西方海上交流与贸易日益增多。
“马尼拉大帆船”贸易航线与明代外贸
南澳一号

禁海
与朝贡贸易
MARITIME EMBARGOING AND TRIBUTE TRADE

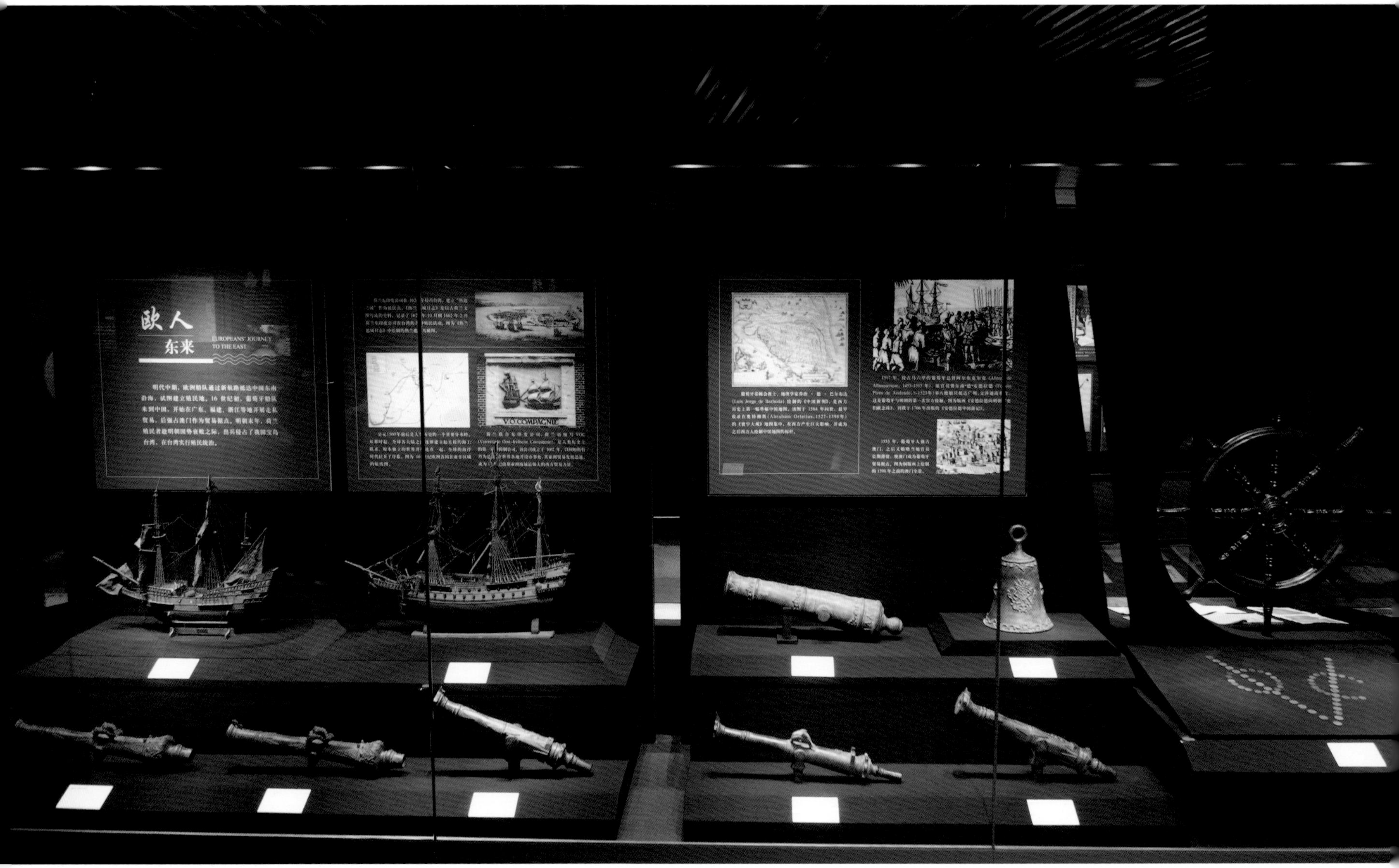
欧人
东来
EUROPEANS' JOURNEY
TO THE EAST

廣州羊城八景圖

粤绣广州羊城八景图（清代）

Eight Sights in Guangzhou of Guangdong Embroidery (Qing Dynasty)

轮船招商局
1866 年，闽浙总督左宗棠创办福州船政局，后由继任船政大臣沈葆桢主持。船政局主要由铁厂、船厂和船政学堂三部分组成，是中国近代史上第一个制造轮船的专业工厂。图为 1868 年福州船政局旧照。
北洋水师大沽船坞
江南机器制造总局
1865 年，江南机器制造总局成立，它是晚清最大的近代化军事工厂，曾国藩规划，李鸿章实际负责。先后建有十几个分厂，雇用工人 2800 余人，能够制造枪炮、弹药、轮船、机器，还设有翻译馆、广方言馆等文化教育机构。

维护
海洋权益的努力
EFFORTS IN MAINTAINING THE
MARINE RIGHTS AND INTERESTS
海洋强军建设
萨镇冰
刘冠雄
清末至民国的海权维护
「收复东沙群岛」
「巡视西沙群岛」
张人骏

中华海洋文明

第三篇章

Marine Civilization of China 3

中华海洋文明第三篇章分为认知海洋、人海和谐、依海富国、以海强国、合作共赢五部分，全面展示了中华人民共和国成立后，中国依托蔚蓝海域从海洋大国走向海洋强国的宏伟历程。

Marine Civilization of China 3 is divided into five parts: "knowing the ocean", "harmony between people and cean", "prospering the country by ocean", "strengthening the country by ocean", and "win-win cooperation", which fully demonstrate the magnificent course of China from a big maritime country to a strong maritime country since the establishment of the People's Republic of China, relying on the blue ocean.

前 言

中国是一个发展中的海洋大国，拥有广泛的海洋战略利益。建设海洋强国是中国特色社会主义事业的重要组成部分。新中国成立以来，特别是改革开放以来，我国的海洋事业不断发展壮大，从第一批沿海开放城市，到21世纪海上丝绸之路；从耕海牧渔、油气开发，到极地大洋科考、海洋生态文明建设。在习近平新时代中国特色社会主义思想的指引下，我国坚持陆海统筹，坚持走依海富国、以海强国、人海和谐、合作共赢的发展道路，通过和平、发展、合作、共赢方式，扎实推进海洋强国建设。

China is a developing maritime power with wide range of maritime strategic interests. Building a powerful maritime country is an important part of the cause of socialism with Chinese characteristics. Since the founding of the People's Republic of China, especially since the reform and opening up, Chinese marine industry has been growing significantly, from the first batch of coastal open cities to the 21st Century Maritime Silk Road; from marine farming, fishing, oil and gas exploitation to polar ocean scientific research and marine ecological civilization construction. Under the guidance of Xi Jinping's thought of socialism with Chinese characteristics in the new era, China adheres to the principle of land and sea integration, adheres to the development path of enriching the country by relying on the sea, strengthening the country by the sea, achieve the goals of harmony between human and sea, and solidly promotes the construction of a powerful marine country through peaceful, positive, cooperative and win-win ways.

1956

1949

4月23日，人民海军在江苏泰州白马庙应运而生。图为1949年11月8日组建的华东军区海军第一、第二舰大队。

1949

1950

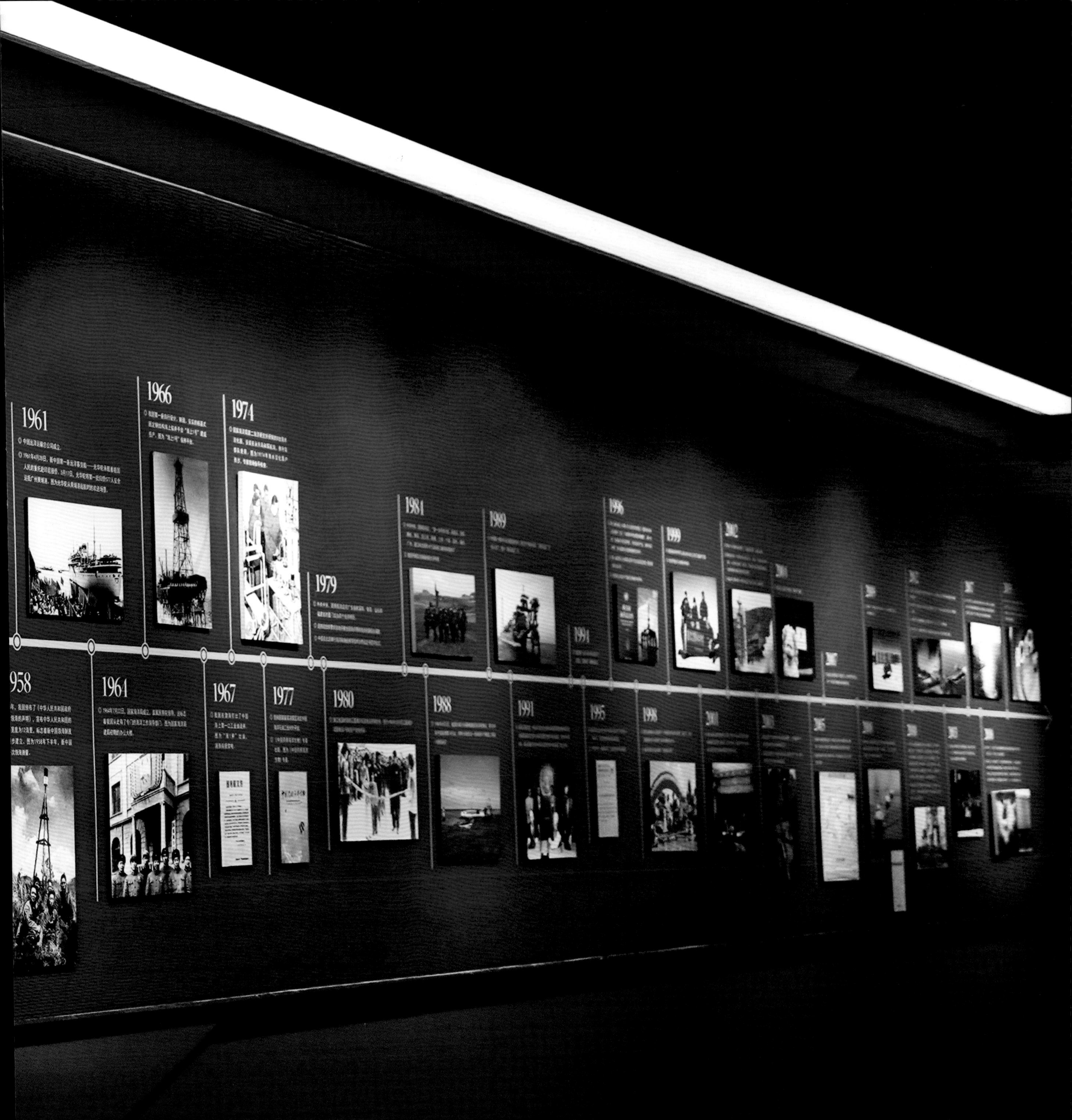

1961
1966
1974
1979
1981
1989
1994
1996
1999
2002
958
1961
1967
1977
1980
1988
1991
1995
1998

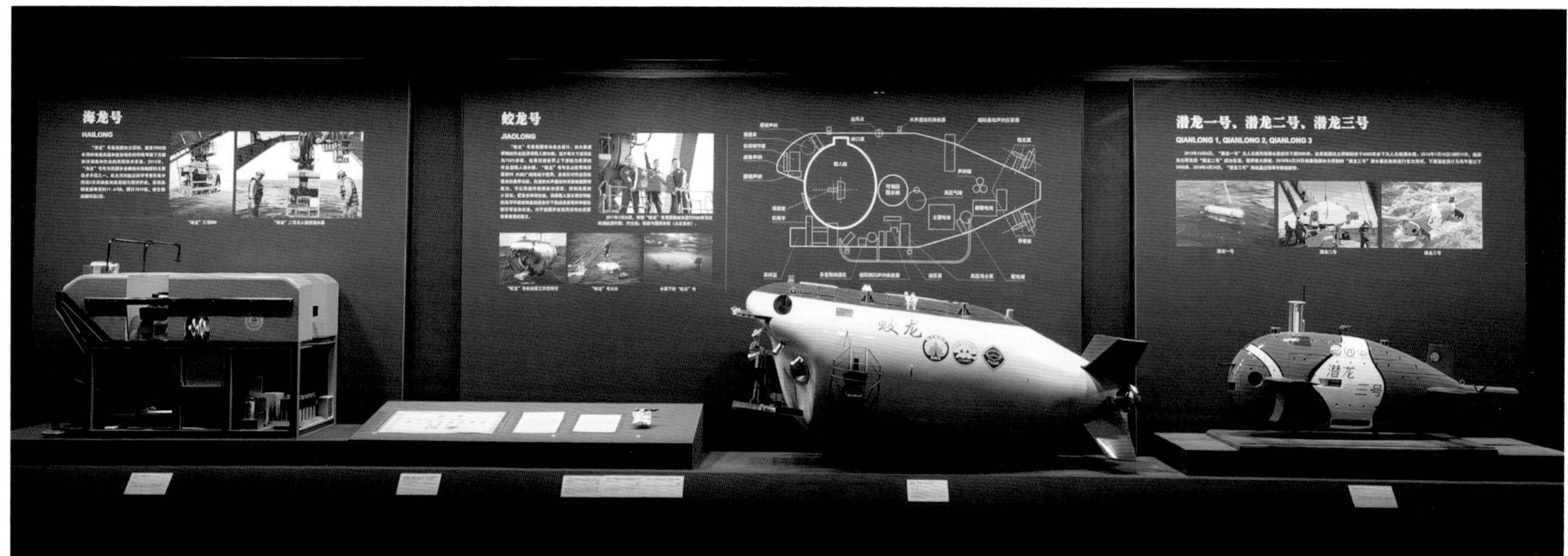

海龙号
HAILONG
蛟龙号
JIAOLONG
潜龙一号、潜龙二号、潜龙三号
QIANLONG 1, QIANLONG 2, QIANLONG 3
蛟龙
潜龙
三号

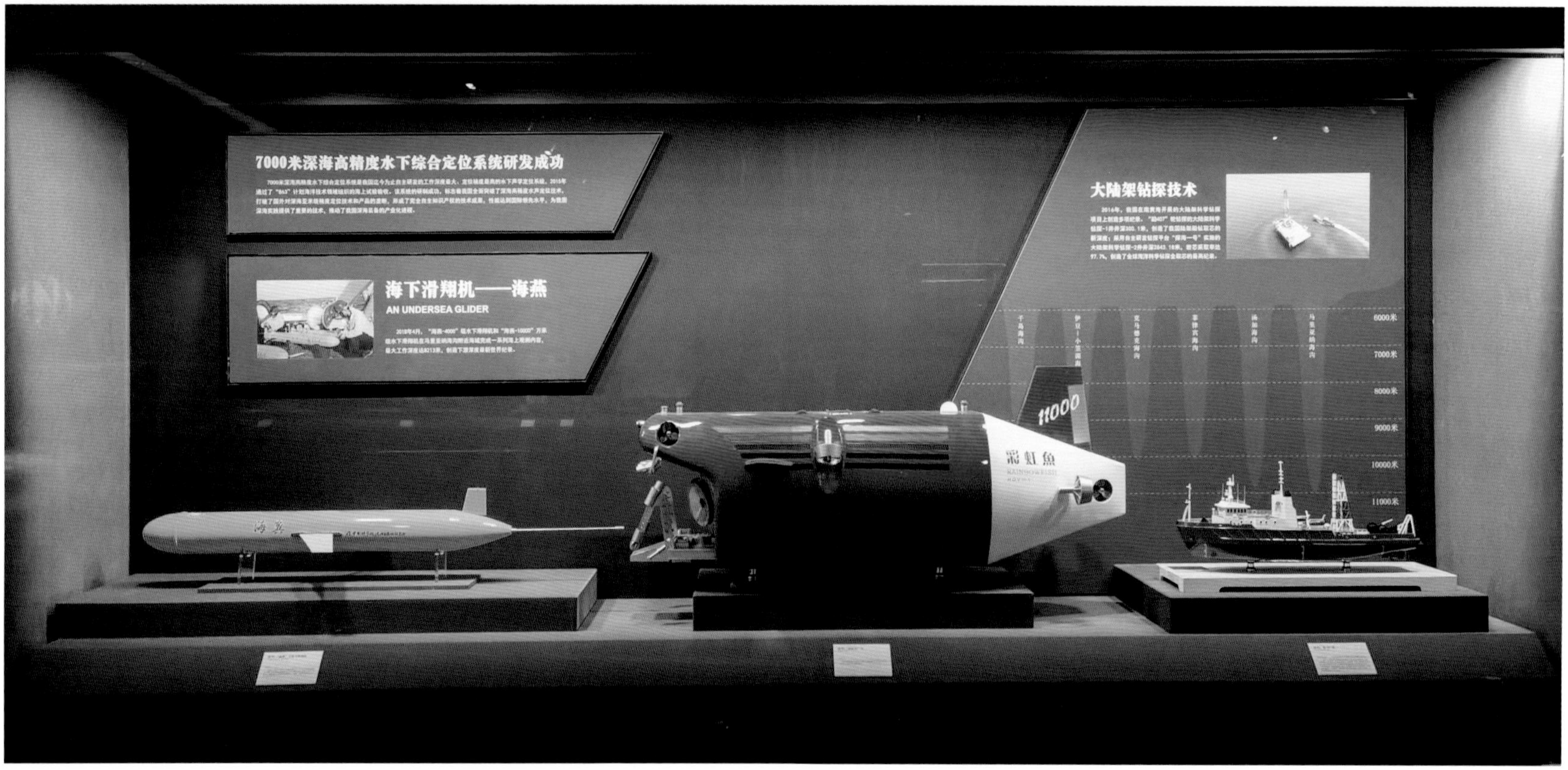
7000米深海高精度水下综合定位系统研发成功
海下滑翔机——海燕
AN UNDERSEA GLIDER
大陆架钻探技术
11000
彩虹鱼
6000米
7000米
8000米
9000米
10000米
11000米

和平利用南北极
PEACEFUL USE OF THE NORTH AND SOUTH POLES
南极考察航线
北极考察航线

索的脚步从未停息
掌闻上海 祝中国第30次南极科考圆满成功
www.xhzw.cn 2013.11.8上线
泰山站建设队

7000米深海高精度水下综合定位系统研发成功
水下滑翔机——海燕
4500 米
感应器
感应器
感应器
感应器
感应器
感应器

“蓝 一号”外形庞大 全长117米 宽度92.7米 高度118米

中央军委在南海海域隆重举行海上阅兵

维护海洋权益与安全

习近平总书记指出，要维护国家海洋权益，着力推动海洋维权向统筹兼顾型转变。我们爱好和平，坚持走和平发展道路，但决不能放弃正当权益，更不能牺牲国家核心利益。要统筹维稳和维权两个大局，坚持维护国家主权、安全、发展利益相统一，维护海洋权益和提升综合国力相匹配。要坚持用和平方式、谈判方式解决争端，努力维护和平稳定。要做好应对各种复杂局面的准备，提高海洋维权能力，坚决维护我国海洋权益。要坚持“主权属我、搁置争议、共同开发”的方针，推进互利友好合作，寻求和扩大共同利益的汇合点。

2012年9月以来，我国积极开展钓鱼岛维权行动，图为中国海警舰船编队在我钓鱼岛领海内常态化巡航。

2015年5月24日至25日，三沙市综合执法1号船完成对南海中沙海域的巡航执法检查。图为三沙市综合执法1号船准备从三沙永兴岛出发巡航执法。

2010年4月，中国海监总队组织海监83船、81船和海监B-7112直升机组成海上编队，执行南海海域专项巡航任务。海上编队对黄岩岛、南海南部、西沙西部、北部湾湾口等海域进行了维权巡航，在曾母暗沙，举行了庄重的中国主权碑投放仪式，宣示国家主权，体现对南海的实际管辖。这是中国海监开展定期巡航执法以来首次投放主权碑，影响深远，具有重要的历史意义。

2010年4月20日，国家海洋局在“中国海监83”船前甲板举行南海巡航曾母暗沙投碑仪式。

2014年，海军南海舰队远海训练编队巡航曾母暗沙。

2013年3月26日，中国渔政46012船赴西沙及黄岩岛海域巡航护渔。

2017年4月，万吨级中国海警3901船在西沙海域开展海域海岛保护联合执法。

远航时代
永无止境的探索之旅

The Age of Oceangoing Voyage
A Never-ending Journey of Exploration

广袤与未知，是海洋和星空共同的特征。
海洋曾以宽广隔绝大陆，
却因人类的探索连缀成新的世界，
从此需求重计疆域，物种互通有无。
而今，人类的视线投向更加辽远无垠的天际。
从一个远方，到另一个远方，
我们又将起航，目标是遥不可知的星系，是亿万光年外的神秘，
更是关于过去、现在与未来的永恒追索。

Being vast and unknown is the common feature shared by the ocean and the sky.
The ocean once isolated the continents with its broadness,
but the continents were connected to a new world because of human exploration.
From then on, the demand has come beyond territories, and the species have been shared and exchanged.
Nowadays, human eyes are cast on a more distant horizon.
From one faraway place, to another faraway place,
we will set sail again, targeting distant unknown galaxies, and discovering mysteries of hundreds of millions of light-years away.
It is more of an eternal search for the past, present and future.

航海发现之旅
展厅

Voyages of Discovery Exhibition Hall

航海发现之旅展厅以西方航海大发现为线索，展现了航路沟通大陆、物种互通有无的壮阔之旅，分为序厅、达尔文的环球考察、探秘大洋洲、冰冻荒原、非洲之旅、深入新大陆、奇境欧亚七部分。

Taking the voyages of discovery in the west as a clue, the Voyages of Discovery Exhibition Hall shows the magnificent journey of the voyages to bridge the continents and exchange species. It is divided into seven parts: "preface hall", "Darwin's global expedition", "the exploration of Oceania", "the frozen wasteland", "the journey to Africa", "going deep into the new world", and "the wonderland of Eurasia".

ENNETH E.BEHRING
肯尼斯·贝林
传奇人生
LEGENDARY LIFE

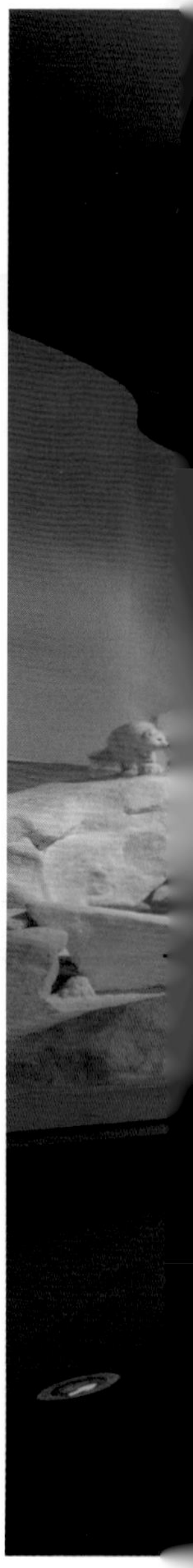

适应奥秘

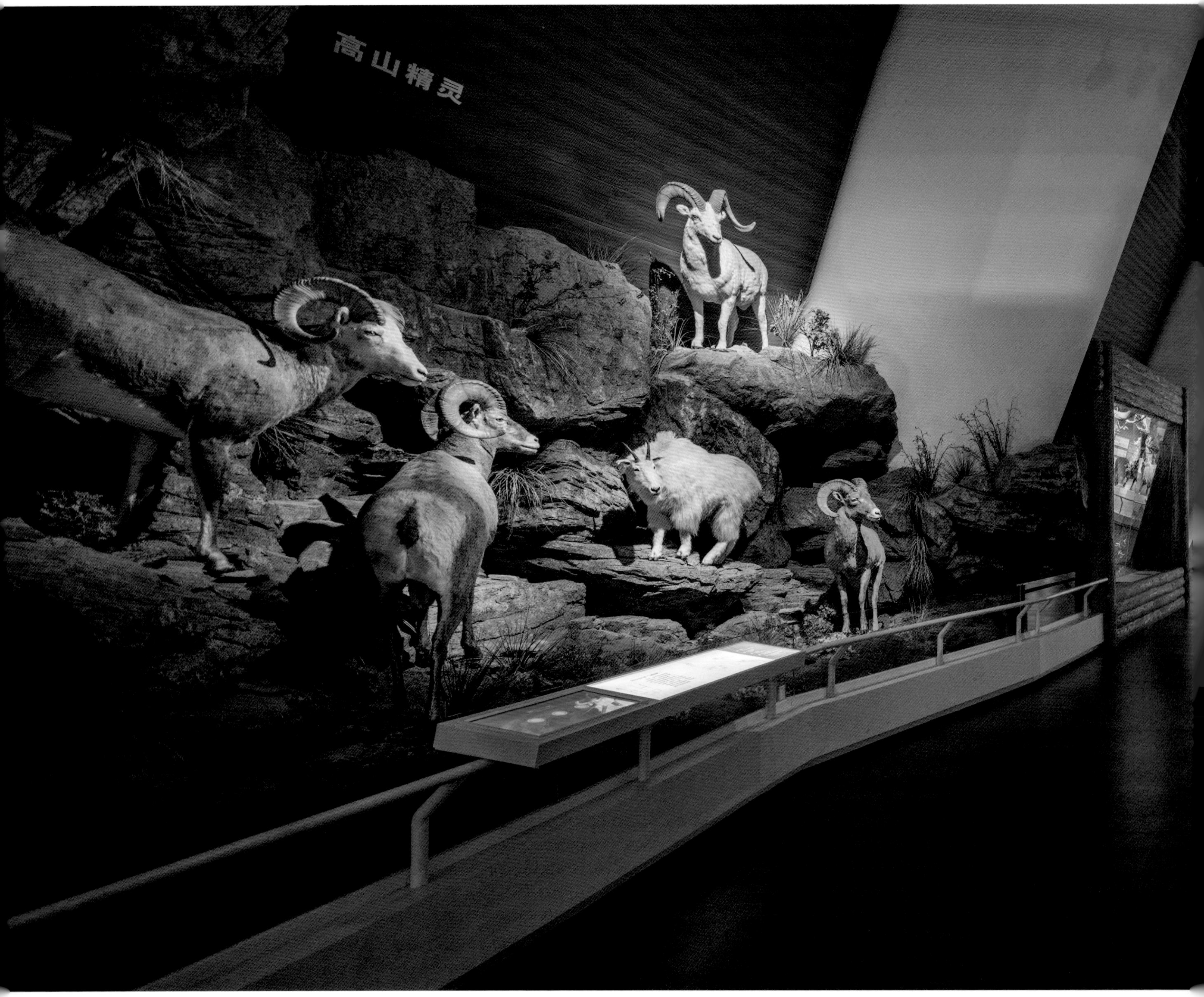

大航海大交换

从风帆到行轮

展厅

From Sailboat to Steamship Exhibition Hall

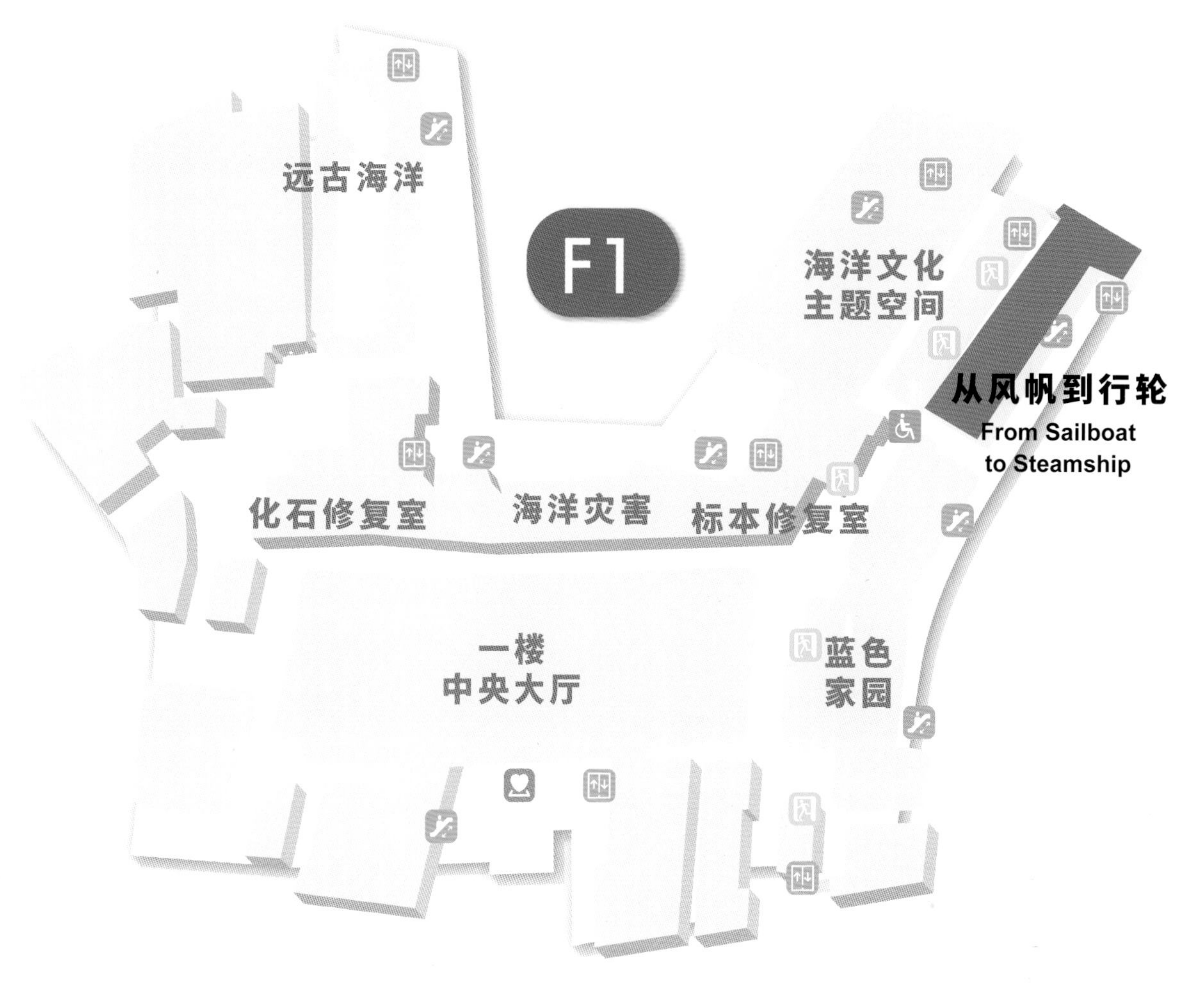

从风帆到行轮展厅以西方近现代船舶制造和航海技术发展简史为背景，以馆藏19世纪至20世纪初西方船舶设备、器具为核心，以国际化的视角展示了近现代航海技术的基本轮廓与发展路径。

Taking a brief history of the western modern shipbuilding and maritime technology development as the background, the From Sailboat to Steamship Exhibition Hall presents the basic outline and development path of modern maritime technology from an international perspective, with a collection of western ship equipment and apparatus from the 19th century to the early 20th century as the core.

帆船时代的终结
The end of the Age of Sail

请勿触摸
NO TOUCHING

船舶驾驶台天花板
顶部甲板
罗经
罗经反射式
观察筒
平面反射镜
为减少驾驶台结构对磁罗经干扰，以及改善驾驶台空间的拥
挤，反射式磁罗经被安装在船舶驾驶台顶部的甲板上，罗经柜中
的镜筒通过罗经柜基座穿过驾驶台天花板向下伸入驾驶室内，驾
驶员通过安装在驾驶台内的镜筒端部的平面反射镜，观察罗盆刻
度盘投影映像，以获取磁罗经相应的航向或方位读数
八分仪

船舶航行灯的显示方式

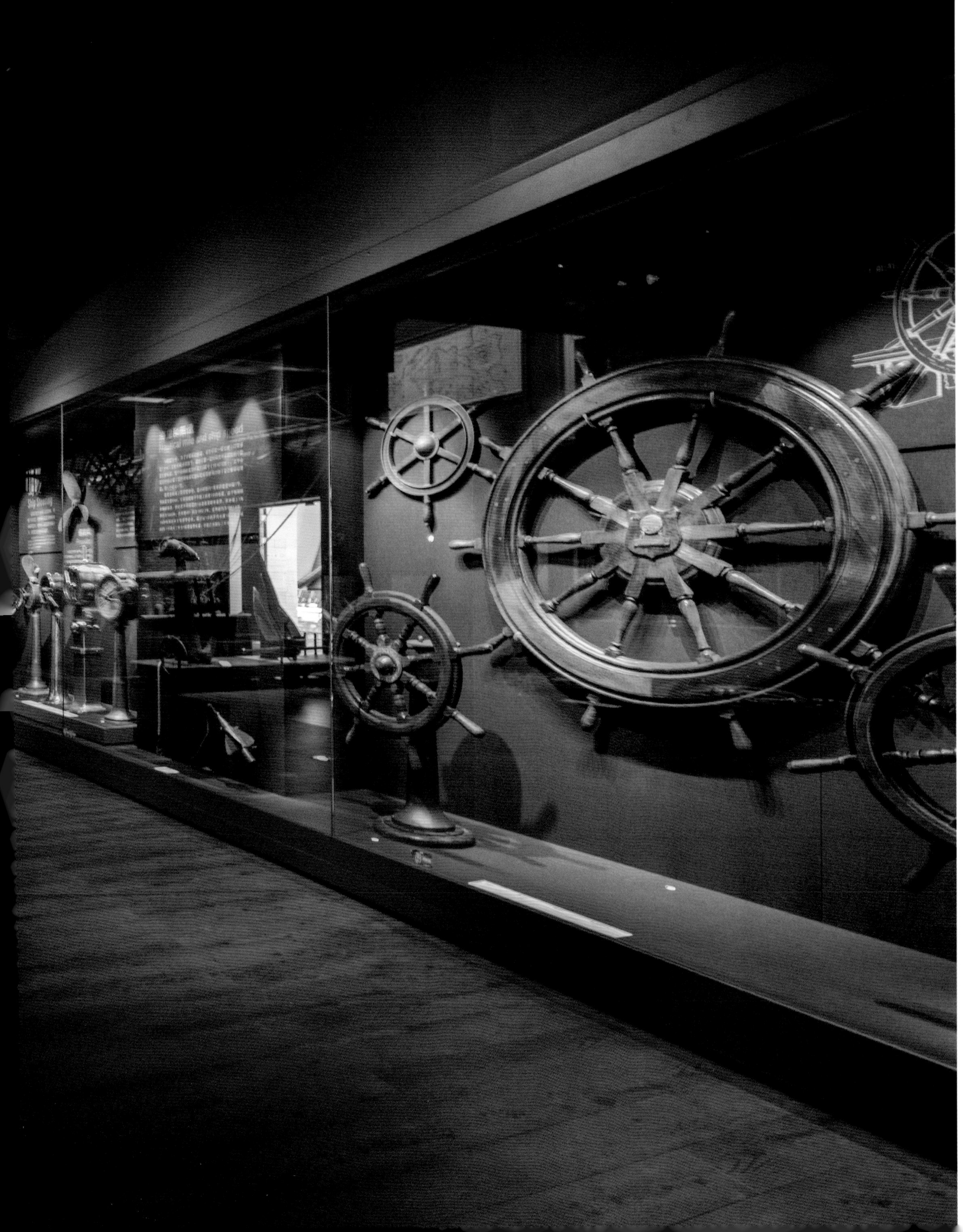

舱内一角
Inside the cabin
舱内设置试图将技术与艺术自然地融为一体，创造出一个舒适的生活空间。
The cabin design tried to integrate technology and art so naturally as to create a comfortable living space.
TOLEDO

筑梦极地
展厅

Dreams of Polar Regions Exhibition Hall

筑梦极地展厅以生命与环境、探索与发现、极地与人类为主题，全面展现了极地重要的气候地位，灵动的生命景象和科学家孜孜探索与勇于挑战的精神。

With the theme of life and environment, exploration and discovery, and polar regions and human beings, the Dreams of Polar Regions Exhibition Hall shows the important climate status of the polar regions, the dynamic life scenes and the spirit of scientists' diligent exploration and courage to challenge.

北极生命
ARCTIC
LIFE

雪 龍
XUE LONG

CHANGES
OF THE POLE 两极的变化
2014/2/8
2018/2/7

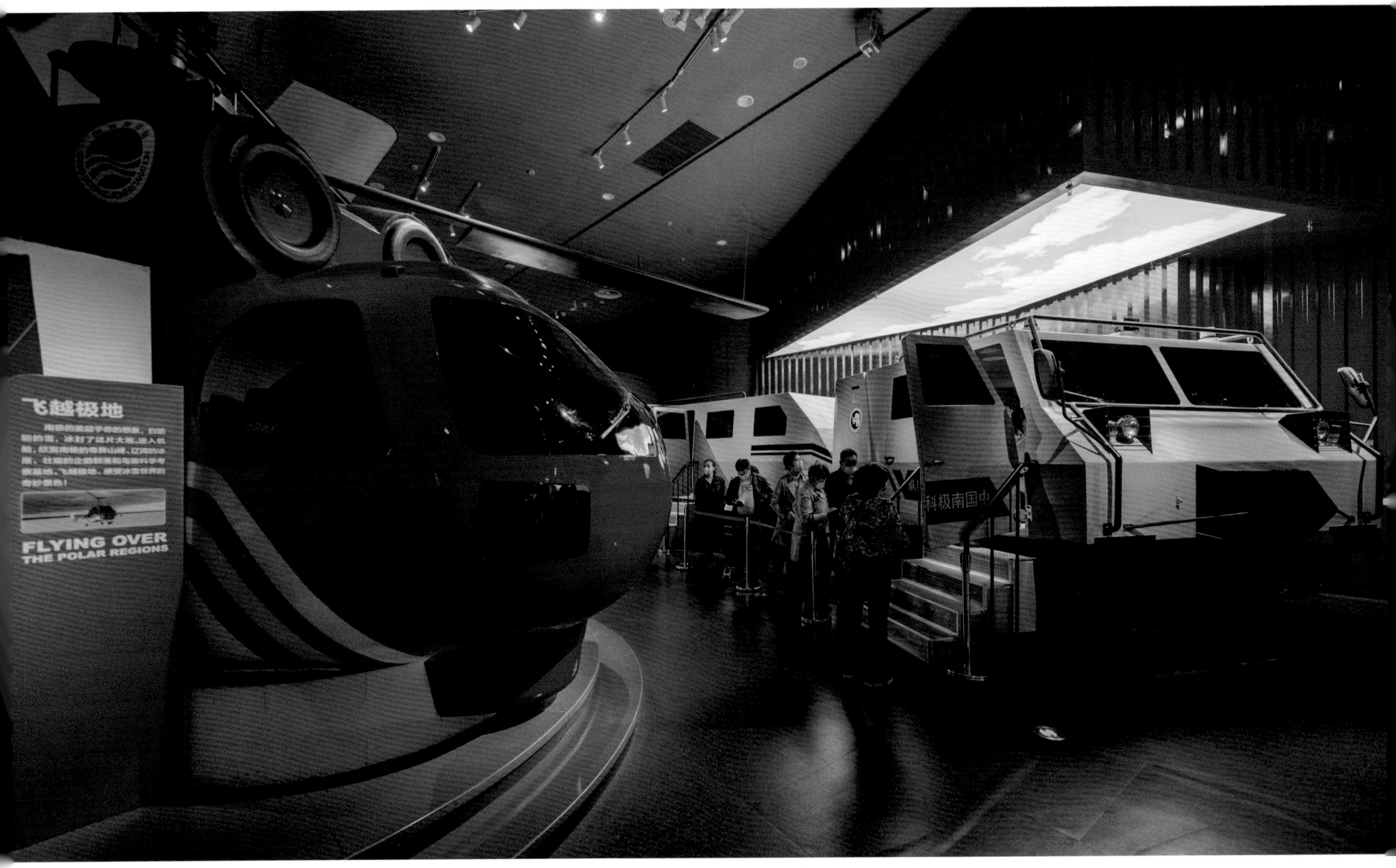
飞越极地
FLYING OVER
THE POLAR REGIONS

国际南极内陆考察路线
INTERNATIONAL EXPEDITION
ROUTE IN THE SOUTH POLE
极地地区的探索史
THE EXPLORATION
OF THE POLAR REGIONS
"雪龙2"号
极地科考破冰船

海洋天文
展厅

Marine Astronomy Exhibition Hall

海洋天文展厅以人类海洋活动中的天文学发现与发明为线索，连缀起海洋与星空、地球与宇宙的宏大叙事，用永无止境的好奇与探索之心，宣誓星辰大海的崭新征程。

The Marine Astronomy Exhibition Hall takes the astronomical discoveries and inventions in human marine activities as clues to connect the grand narratives of the ocean and the sky, and the earth and the universe, swearing a brand-new journey to the sky and the ocean with never-ending curiosity and exploration.

开阳
摇光
开阳增一
玉衡
天权
天枢
天玑

东西方天文探索的对比

Comparison of Astronomical Exploration between East and West

二十八星宿，十二星座，中国与西方用不同的名称标明了同一片星宇，又同样将之用于远航的探索、祸福的想象。这是好奇心的不约而同，也驱动着人类迈向更遥远的未知。

The twenty-eight mansions in China and the twelve constellations of the zodiac in the west—China and the west used different names to refer to the same starry universe and also used them in the exploration of voyages and the imagination of weal and woe. This is a coincidence of curiosity that also drives mankind towards the more distant unknown.

大气层之
SPACE TELESCOPE
"大"显神通
MAGIC POWER OF
全光谱时代
AGE OF FULL SPECTRUM
巨眼观天
EYES OF THE SKY
三种光源
A SOUL WITHOUT IMAGINATION
IS LIKE AN OBSERVATORY
WITHOUT A TELESCOPE.

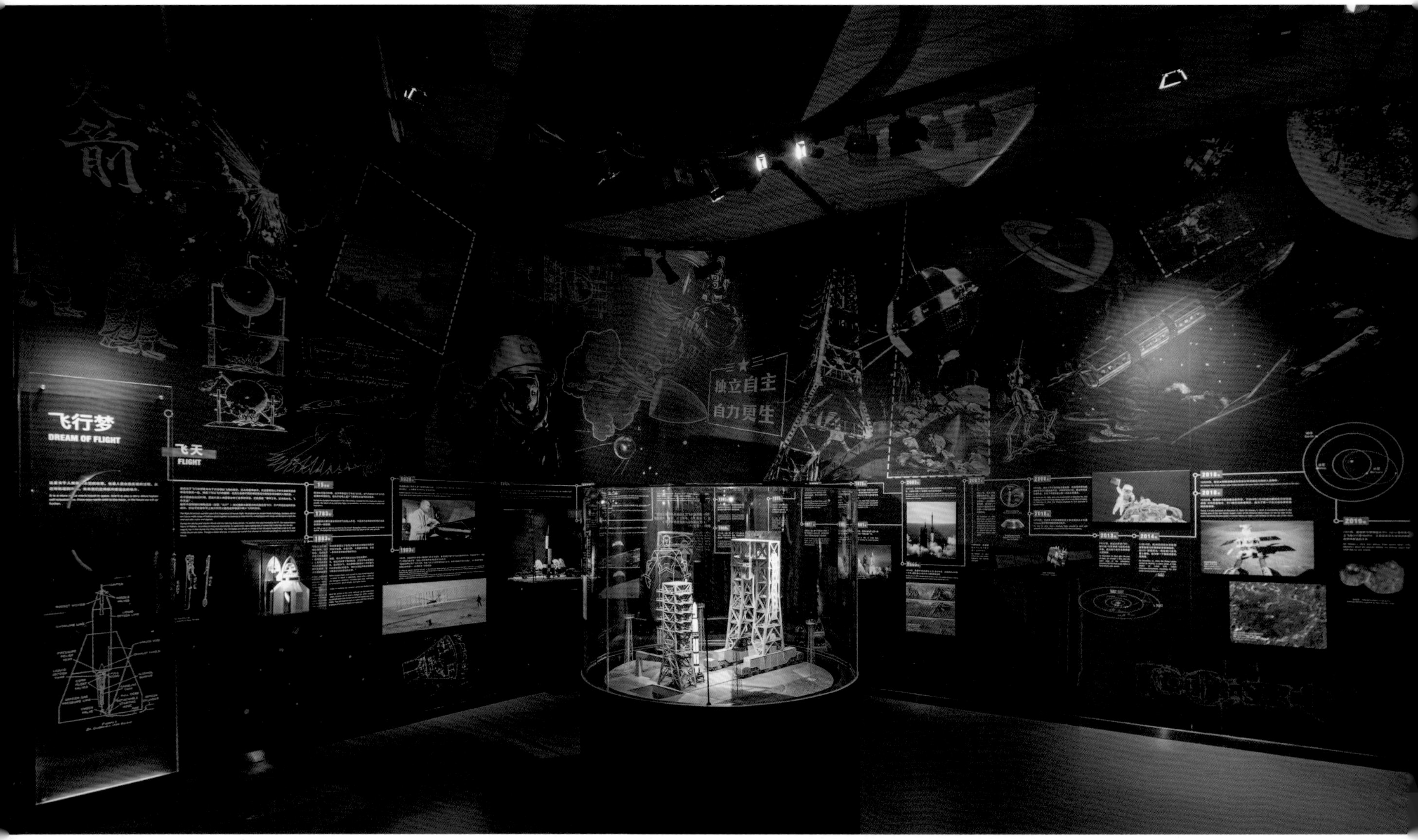
飞行梦
DREAM OF FLIGHT
飞天
FLIGHT
独立自主
自力更生

“大”显神通
MAGIC POWER OF
VARIOUS TELESCOPES
全光谱时代
AGE OF FULL SPECTRUM
巨眼观天
EYES ON THE SKY
三种光路
A SOUL WITHOUT IMAGINATION
IS LIKE AN OBSERVATORY
WITHOUT A TELESCOPE
没有想像力的灵魂
就像没有
望远镜的天文台

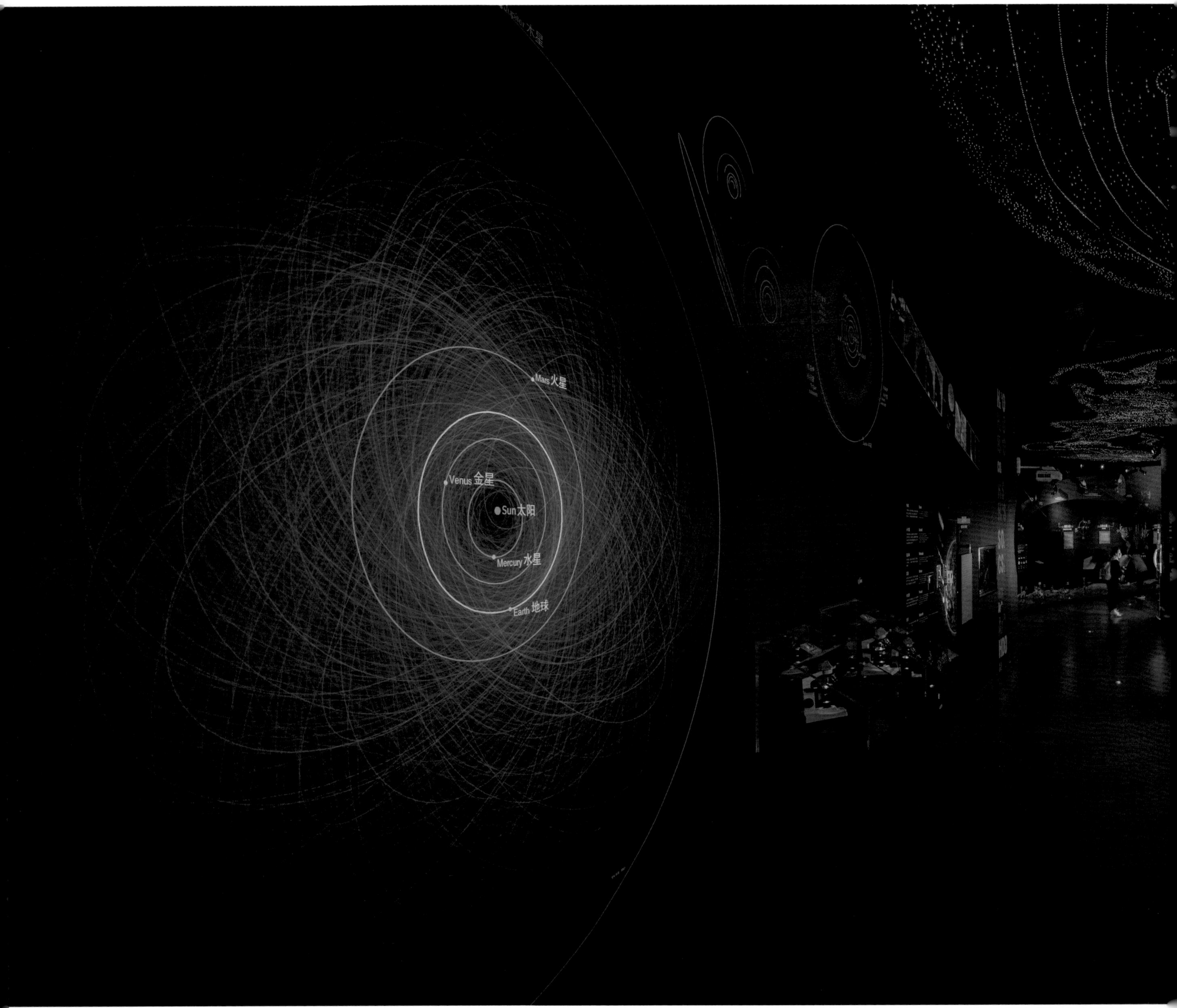
Mars 火星
Venus 金星
Sun 太阳
Mercury 水星
Earth 地球

双子座
Gemini
大犬座
Canis Major
猎户座
Orion

球幕影院 / Dome Theater

1989
海王星
旅行者2号

太阳系八大行星模型场景

Model Scenes of the Eight Planets in the Solar System

乐享海博

文旅融合的魅力图景

Enjoying Yourself in National Maritime Museum of China

Charming Pictures about Culture-tourism Integration

互动体验区

互动体验区的“欢乐海洋”是小朋友的乐园，在以海洋知识为线索的游乐项目中，孩子们可以进一步感受与海有关的奇趣魅力。

Interactive Experience Area

The "Sea of Joy" in the Interactive Experience Area is a paradise for children. In the amusement project with the knowledge of the ocean as a clue, the children can further feel the fabulous charm of the ocean.

科普拓展区

科普拓展区包括对外开放的化石修复室、标本修复室、海博电影院和科普教育中心，在专业、亲和、寓教于乐的氛围中增强青少年的亲海、识海、护海意识，成为海博展览与教育边际的延伸。

化石修复室 / Fossils Repair Room

Popular Science Development Area

The Popular Science Development Area includes a fossils repair room, a specimen restoration room, a cinema of the National Maritime Museum of China and a popular science education center, which are open to the public. In a professional, friendly and entertaining atmosphere, it enhances young people's awareness of loving, knowing and protecting the ocean, and it is an extension of the exhibition and education of the National Maritime Museum of China.

标本修复室 / Specimen Restoration Room

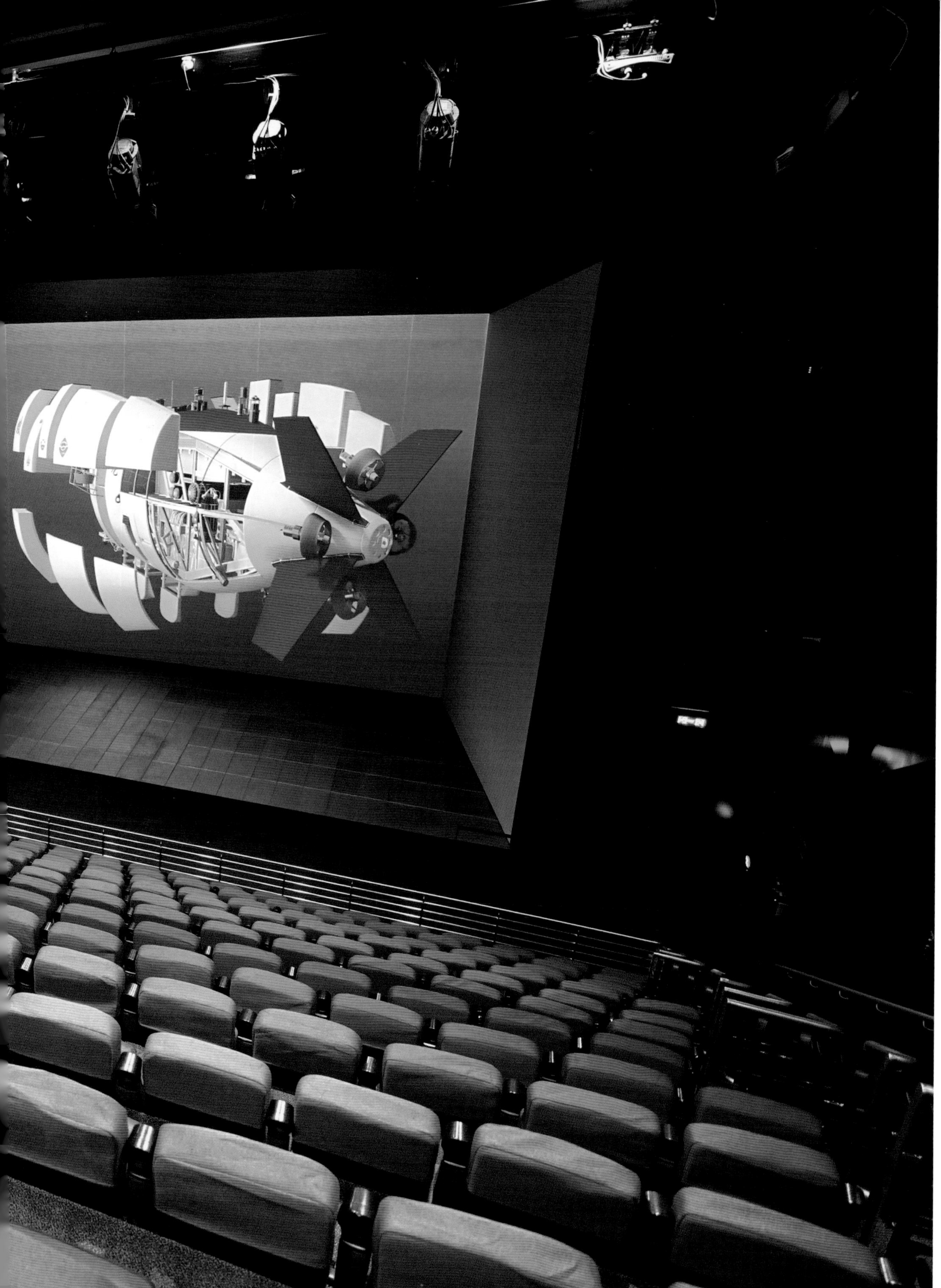

海博电影院

Maritime Museum
Cinema

海洋实验
MARI

科普教育中心 / Popular Science Education Center

科普教育中心 / Popular Science Education Center

海洋生物
MARINE LIFE
威瓦西虫

科普教育中心

Popular Science Education Center

其他功能区

国家海洋博物馆致力全流程智慧博物馆体系建设，在关注智慧管理、智慧保护、智慧服务、智慧运营的同时,不断优化各类游客服务项目，为提升运营效能、改善游客体验奠定了基础。

游客入口闸机 / Visitor Entrance Gate

Other Functional Areas

The National Maritime Museum of China is committed to the construction of a whole-process intelligent museum system. While focusing on intelligent management, intelligent protection, intelligent services, intelligent operation, it constantly optimizes various visitor services, laying a foundation for improving operational efficiency and visitor experience.

导览台 / Guide Service Desk

智能导览机 / Intelligent Audio Guide

冷饮简餐餐厅 / Cold Drinks Dining Room

餐厅 / Restaurant

咖啡书吧 / Cafe & Bookstore

纪念品商店 / Souvenir Shop

EXIT

百年历史的“绿眉船”

There is a century-old "green eyebrow boat"

歼教-6型超声速喷气式歼击教练机
JJ-6 Supersonic Jet Fighter Trainer

“大溪地号”独木舟
"Tahiti" Canoe

“余庆号”导护艇 / "Yuqing" Missile Corvette

后记

国家海洋博物馆运营开放至今，已有两年多时光。

作为我国唯一的国家级综合性海洋博物馆，自立项之日起，国家海洋博物馆的建设进程便备受关注。但由于客观条件所限，在很长一段时间里，她的完整规划构想和策展方案都面临缺位。

2018年2月，时任天津自然博物馆馆长的我接到调令，前往天津市海洋局任副局长，兼任国家海洋博物馆筹建办主任，正式接手国家海洋博物馆的各项规划与建设工作。

我深知，国家海洋博物馆的兴建，是中华文明蓝色史诗的崭新篇章，是中华民族向海而兴的庄严宣告，更是新时代中国人对未来和世界的蔚蓝诺言。兹事体大，交于我手，要在紧迫的时间里打造展览、实现开放，这是压力，是重担，更是信任，是责任。

在前期负责同志和各位专家的调研工作的基础上，我们逐渐确立了国家海洋博物馆的总体规划方案。她将从人类命运共同体的宏大叙事出发，串联地球、生命与人类的脉络，谱写历史、现在与未来的华章。

在具体策展中，我们尝试以多剖面的视角展示海洋与人类息息相关的全貌。展览中呈现的海洋，滋养的是海物维错，也是陆地众生；标志的是乘风破浪，也是物种互通；诉说的是海权荣辱，也是奋进图强。

这样的方案，有幸得到了当时的主管市领导的肯定，并获得了天津市海洋局党组的同意和批准，自然资源部及其所属涉海单位更为我们的展品征集和开馆工作提供了巨大的帮助。在多方支持下，2019年5月1日，国家海洋博物馆完成展陈基本建设工作，顺利对外开放。至此，这场跨越近十年的接力终于暂告圆满。我们要感谢的，是长期以来为国家海洋博物馆建设付出的每一个人。正是他们，用功成不必在我、功成必定有我的担当，推动了国家海洋博物馆的顺利启航。

国家海洋博物馆的开放不是终点，而是另一个起点。短短一年有余的时间，展陈策划必定难以尽善。在运营过程中，我们还将根据各方建议和反馈，不断进行调整、优化和升级。我相信，在各界有识之士的共同努力下，国家海洋博物馆必将成为中华民族伟大复兴进程中的一个璀璨节点，向世界讲述属于中国、造福人类的海洋故事，让中国的海洋声音响彻寰宇。

Postscript

It has been more than two years since the National Maritime Museum of China was open to the public.

As the only national comprehensive marine museum in China, the construction process of the museum has attracted much attention since the day the project was established. However, limited by objective conditions, the museum faced the absence of the completed planning concept and curatorial scheme for a long time.

In February 2018, I, then curator of the Tianjin Natural History Museum, received a transfer order to the Tianjin Oceanic Bureau as deputy director and concurrently served as the director of the Preparatory Office of the National Maritime Museum of China, officially taking over the planning and construction work of the museum.

I'm well aware that the construction of the National Maritime Museum of China is a brand-new chapter in the blue epic of Chinese civilization, a solemn declaration of the rise of the Chinese nation to the ocean, and a blue promise of Chinese people to the future and the world in the new era. Facing the tight deadline, I undertook this heavy but quite important task to plan an exhibition and open the museum to the public. Though under great pressure, I shouldered the responsibility because of the trust I had received.

Based on the research of comrades and experts in the early stage, we gradually established the master plan of the National Maritime Museum of China. Starting from the grand narrative of a community with a shared future for mankind, the museum could link the veins of the earth, life and human beings, and write a brilliant chapter of history, present and future.

In the specific curation, we tried to show the whole picture of the ocean's relevance to human beings from a multi-faceted perspective. The ocean presented in the exhibition nourishes the sea creatures and land beings; it symbolizes the braving wind and the waves and the interchange of species; it speaks of the glory and disgrace of maritime power and the progress and strength.

Such a plan was fortunately affirmed by the then municipal leaders, agreed and approved by the leading CPC group of Tianjin Oceanic Bureau, and received strong assistance from the Ministry of Natural Resources and its subordinate sea-related units for the collection of exhibits and the opening of the museum. With the support of many parties, on May 1st, 2019, the National Maritime Museum of China completed the basic preparatory work of the exhibition and was opened to the public successfully. On that day, the relay spanning nearly a decade has finally come to a successful conclusion. We would like to thank everyone who has contributed to the construction of the National Maritime Museum of China for a long time. It is their selfless contributions that promote the smooth opening of the National Maritime Museum of China.

The opening of the museum is not the end, but another starting point. In just over a year's preparation, it's difficult for the exhibition to be perfect. During the operations, we will continue to adjust, optimize and upgrade the exhibition according to the suggestions and feedback from all parties. I believe that, with the joint efforts of insightful people from all walks of life, the National Maritime Museum of China will become a brilliant milestone in the process of the great rejuvenation of the Chinese nation, telling the world the maritime stories that belong to China and benefit mankind, and making China's maritime voice resound throughout the world.